THE
UNAPPROACHABLE
Norton

Murray McLeod

Born in Sydney, Australia the author is an accredited artist specialising in landscape, aviation and portrait subjects. For a number of years Murray and wife Aileen owned and operated their art gallery on Mount Tamborine, a spectacular tourist area on the Gold Coast hinterland of south-east Queensland. He has been a regular contributor to an international online magazine with his aviation and motorcycle articles. Murray also writes and illustrates on a monthly basis for 'Global Aviator' an emerging publication based in South Africa.

Murray was well acquainted with the early post-war racing scene, not as a racer but a keen clubman, touring rider and flag marshal at race meetings. His riding career paralleled the inaugural World Championship era when British riders and machinery greatly influenced the racing calendar.

Today's riders are largely unaware of the sheer magnitude of the British industry during those pre-war and early post-war years. With twenty or so manufacturers offering a comprehensive range of models, one was rather spoilt for choice. At one time or other the author's stable was graced by leading makes; Norton, Velocette, Vincent, Triumph and Matchless to name just a few. The industry's high point was probably reached around 1955 and yet by 1975 it had virtually ceased to exist. The same could be said for British race machinery, overwhelmed firstly by Italian dominance which in turn suffered the same fate by products from 'The Land of the Rising Sun'.

With these events the writer's abiding interest in road racing tended to wane somewhat. However the tradition is still maintained by way of Historic Road Racing where veteran riders campaign their precious machinery with the same old enthusiasm. How better to perpetuate their memory than to present a review of their past exploits. This volume represents years of enjoyable research and provides an insight into those exciting eras when British bikes ruled the circuits, in particular the iconic Norton and the larger than life characters who rode them.

Australian author Murray McLeod is also an accredited artist/illustrator with several publications to his name. These are focussed on aviation history plus two motorcycle titles, 'TT Legends' and 'The Unapproachable Norton', covering road racing from the 1920's to the exciting post war period up to the 1960's. Another of his publications, and one that covers a vastly different arena is 'Aussie Tennis Legends', an appreciation of the esteem in which Australia was held over many decades of international and Davis Cup participation.

Author Titles

Nonfiction
Aces and Adventurers
Aussie Tennis Greats
Flying Matilda
For Valour
Images of Eagles
The Unapproachable Norton
TT Legends

Fiction
Elliot's Odyssey
The Pilgrimage

Murray McLeod's books are available on amazon.com, through online bookstores and retailers.

Email: mcleodart@westnet.com.au
http://www.mcleodart.com.au

Cover Design by Murray McLeod
Document creation: Linda Ruth Brooks
www.lindaruthbrooks.com

Photographic acknowledgement:
The Motor Cycle
Motorcycling
Keig Collection
Famous Racing Motorcycles (Griffith)
Original Artwork © by the Author

ISBN-13: 978-1477659724
ISBN-10: 1477659722

A copy of this book can be found in the National Library of Australia.

The author has taken all possible care to give appropriate acknowledgement and seek permissions from all interested parties and welcomes any further correspondence. Enquiries should be addressed to the copyright owner.

Dedication

To old club mates, leather-clad and helmeted
they dared to challenge the odds.

Contents

Introduction

Norton

The greatest name in British motorcycle racing history

Profiles of the Norton works riders of the 1930s

In a world highlighted by extreme sports few activities match the demands of motor cycle road racing at its highest level. For over one hundred years a succession of young men has accepted that challenge and by today's standards those early machines were quite primitive in their specification. Suspension was so basic as to be almost non-existent as were any braking systems.

Transmission was generally belt-driven; an arrangement that restricted the machine to a single-speed situation and an obvious handicap was the inability to surmount steep inclines. Some manufacturers did fit pedal power to their machines to give the rider some assistance to conquer those hills. One can deduce that these were pioneering times and coupled with unsealed roads and diabolical conditions it proved to be a demanding exercise for the aspiring road racer.

From the outset Great Britain was able to produce a seemingly endless crop of talented riders. This was achieved despite efforts by politicians to ban motor sport on public roads in mainland England. Some canny types circumvented the ban by creating circuits on private land. A prime example was the Brooklands circuit at Weybridge in Surrey. It was opened in 1907 and featured vast areas of concrete with incredibly steep banking. The sheer magnitude of Brooklands was more suited for motor racing than for the two-wheeled brigade. Nevertheless from its opening it was strongly patronized by both camps until its permanent closure at the outbreak of war in 1939. While the British Government maintained its aversion to motor racing on public roads, the Isle of Man Parliament did not share their antipathy. From the outset it recognized the value of motor sport, both as a spectacle and an incentive to tourism and in 1907 the inaugural Tourist Trophy races were launched.

The title 'Tourist Trophy' was significant, for the entrants were using touring machines rather than specialized track racers. Two categories were featured in the race, one for single-cylinder machines and one for those of twin-cylinder configuration.

Twenty-five gallant souls faced the starter for that epic 158 miles event. Not unexpectedly the attrition rate was fairly high. Fifteen machines succumbed to the gruelling conditions and at the finish two makes put their names to a distinguished trophy. Charles Collier on a Matchless won the single-cylinder class at an average speed of 38.22 mph. In the twin cylinder class Rem Fowler brought his Norton home at a respectable 36.22 mph. Charlie Collier and brother Harry went on to become leading British manufacturers with their 'Matchless' brand of motorcycles. James Norton followed a similar path with the machines that bore his name. He was a firm advocate that 'racing improved the breed' and pursued that policy until his untimely death in 1925 at the early age of 56.

James Norton adopted the heading 'Unapproachable' for his machines which seems an unlikely adjective for a slogan. For whatever reason James Norton obviously fancied such a title and from then on it became 'The Unapproachable Norton'.

That humble event of 1907 was the genesis of the incomparable Isle of Man Tourist Trophy series. Its format underwent huge changes over the ensuing decades to the stage where the machines merely bore a passing resemblance to the average touring motorcycle of that era. But through it all, at least until the early 1960s Norton was a force to be reckoned with. Its hey-day was undoubtedly the 1930s and to a further extent the early post war years and into the 1950s. Such was the dominance of British riders on Norton 'works' machines in that heady 1930 period that in their continental races it was possible for the team manager to pre-arrange the winning order.

More than any other it was Norton that kept British motorcycles in the forefront of international road racing. It was a period when the sole restriction was the cubic capacity of the engine. Supercharging, or forced induction was quite permissible although British designers regarded it as an unnecessary complication. AJS, Velocette and Vincent HRD made tentative forays into that system but the results did not justify the effort. On the other hand Germany and Italy had created multi-cylinder designs that were amenable to forced induction. By 1937 the blown DKW and flat-twin BMW had achieved a degree of speed and reliability to outpace the outmoded single-cylinder British designs. Despite heroic rides their 'works' riders, Norton were forced to concede their previous mantle of success. BMW's star rider Georg Meier was crowned European Champion of 1938 but even more galling to British pride was a BMW 1-2 in the 1939 Isle of Man Senior TT, an all-foreign victory on one's home soil!

During the 1939 season, BMW was usurped by Italian star Dorino Serafini on the supercharged Gilera four. Serafini went on to become European Champion. A memorable victory was his win in the 500cc Ulster TT on the demanding Clady circuit. But more disturbing events were emerging—September 1939 saw the outbreak of World War 2. Eight long years were to elapse before the Tourist Trophy series re-emerged. Just two members of that illustrious Norton team, Daniell and Frith resumed their pre-war career and in both instances they were singularly successful.

Here we have a two-part revue of these giants of the Grand Prix arena, talented riders at the height of their prowess in the cut and thrust of road racing. By nature each one was generally of an independent nature. Some, like Stanley Woods and Walter Handley were not amenable to team orders and eventually went their various ways. One name that stood apart from the others was the quiet Scot from Hawick the immortal Jimmie Guthrie. With his passing came other greats, Harold Daniell and Freddie Frith, whose careers continued into the post war period. By the late 1980s just one Norton team member from that great era was still alive, that person was John White the studious schoolmaster who earned the soubriquet of 'Crasher' before he joined the team. It proved to be an unwarranted title, for in his five years with Norton John White proved to be a most reliable team man who rarely failed to finish and also won several major European Grands Prix.

As the 1930 decade emerged, motor racing was facing a fairly bleak future. By then the repercussions of the Wall Street crash had extended to Europe and the British Isles. Unemployment was soaring to record levels, businesses were closing their doors and particularly hard hit were motorcar and motorcycle manufacturers. Many famous names were to disappear forever, unable to cope with falling demand. Even before these Depression years Nortons were never a high volume manufacturer. Their average annual output rarely exceeded 5000 units which barely match the weekly production figures of current Japanese factories.

Yet through that barren period Norton fielded works teams to contest the TT races and the major classics. Their involvement was only made possible by trade support from petrol and oil giants like Shell and Castrol. The components industry also gave generous assistance, Lucas magnetos and Renold chains were typical examples. Most riders' incomes came from these trade sources, bolstered by start money and funds from race winnings. Compared to the millionaire status of today's top-flight riders it seems barely credible that riders of the 1930s received such paltry incomes from their employer. Nevertheless a win in the TT races, supplemented with various trade bonuses could possibly net a rider two thousand pounds, a considerable sum in those days. However, only a handful of riders could hope to attain that goal and then only as a works entrant. Besides engaging the best riders the driving forces behind Norton's near-invincible situation were dependent on two pertinent factors.

In the late 1920s the path to race success was deemed to be by way of overhead camshafts. Velocette and to a lesser extent Chater Lea were among the first to initiate the trend. From the outset the K Series Velocette was a winner. Alec Bennett took one to a record-breaking win in the 1926 Junior TT; a feat he repeated in 1928 plus a win by Freddie Hicks in 1929. Norton's response was Walter Moore's CS1 which was by his admission an unabashed copy of the Chater Lea. The versatile Alec Bennett rode a CS1 to victory in the 1927 Senior TT but the Sunbeam and Rudge of that era, both of which retained pushrod operation of the valves, generally outclassed the CS1.

In 1929 Walter Moore was enticed away from Norton to become chief designer for the German NSU firm. Not surprisingly his first products bore striking similarities to the original CS1 but Moore's departure from Norton made way for the innovative Arthur Carroll and his new overhead camshaft design. It was a winner from the start; a simple and rugged motor that was amenable to painstaking development over the years. Production of the camshaft Norton was terminated in 1962 and although it was developed to its ultimate form the origins of Arthur Carroll's original design were still discernible.

The other factor in Norton's success was the appointment of Joe Craig as technical director and race manager. A native of Ballymena he rode Nortons with considerable success in his native Ireland and to a lesser extent in the Isle of Man. Joe Craig retired from racing in 1929 and began an involvement with Norton that endured until his retirement in 1955. During World War 2 he spent a period with BSA and later with AJS. With the return of racing in 1946 he returned to Norton and resumed a career interrupted by the war.

Much has been written about Joe Craig's personality and his relationship with the works riders. Some people, such as Stanley Woods remembered him with affection while others were not so charitable. The fact remains that Joe Craig was an astute judge of a rider's abilities. He also gained a reputation for his single-minded adherence to the basic Norton design. This attitude continued through the1930s and until the firm's official withdrawal from Grand Prix racing in 1955.

During that period Craig took the single cylinder Norton to its ultimate stage of development. Alas, it could no longer compete with the current Italian designs. Gilera and MV dominated the 500cc class with their four cylinder 'fire engines'. Even more innovative were Moto Guzzi and their ultra-light 'Gambalunga' series. They won a host of world championships in both the 250 and 350 classes and in 1953 to demonstrate their versatility they revealed an in-line four plus the fabulous 500cc water-cooled V8 in 1956. Then at the end of the 1957 season the leading Italian factories announced their withdrawal from racing. It was a mutual decision they claimed, due to the unsustainable costs of Grand Prix racing. As a result, Moto Guzzi, Gilera, Mondial and others quietly departed the scene. Not so MV, who somewhat brazenly re-appeared in 1958 to begin an era of undisputed domination? They were also blessed with a succession of top-line riders; Surtees, Hocking, Hailwood, Agostini and Read; names that kept MV in the top echelon of road racing until the total dominance of the Japanese. But that, as they say is another story.

The following pages feature profiles of those titans of the 1930s, the pre-war works riders: Harold Daniell, Johnnie Duncan, Freddie Frith, Jimmie Guthrie Walter Handley, Tim Hunt, Walter Rusk, Jimmie Simpson, John White and Stanley Woods. Also featured is Harold Daniell's absorbing commentary of his method of negotiating the challenging TT course where in 1938 he established a lap record of 91mph. This figure would remain unbroken until the1950 TT series when the phenomenon that was Geoff Duke finally eclipsed it in the process of winning that year's Senior TT

Part One

Harold Daniell (1937-1939)

Harold Daniell

1938 Senior TT

Londoner Harold Daniell's racing career was remarkable in that it began in the early 1930s and continued into the post war era. His initial racing forays began at English circuits, such as Donington Park and Syston Park. Daniell was fortunate in having his Nortons prepared by his brother in law, Steve Lancefield who was also a capable rider, however his real talents lay in race tuning.

They made quite a formidable duo on the race circuits and Harold became a consistent winner at Crystal Palace and other venues. In 1932 Daniell was runner-up in the Senior Manx Grand Prix. The Manx races were held in the Isle of Man each September and were restricted to amateur riders. A win in the Manx invariably led to an invitation to join a works team. Daniell's perseverance was rewarded with victory in the 1933 Senior, however the magical invitation from Joe Craig was not forthcoming. Instead Harold became an AJS team member from 1934 to 1936 which proved to be a rather bleak period for the bespectacled Londoner.

AJS were developing a supercharged 500cc V4 that was shaping up as a liability rather than a winner. Their 350 'R' series single was only slightly more reliable and gave Harold just one T.T. finish from his six AJS starts with an eighth place in the 1935 Junior. His AJS tenure had been most unrewarding and in 1937 he was entered for the TT on Lancefield-prepared Nortons. This was a most gratifying exercise for them both with Harold finishing in fifth place in both the Junior and Senior events, the first private entrant behind the works teams.

Joe Craig could scarcely continue to ignore his abilities and later in the season he was recruited into the works team. Harold's first appearance was in the 500cc class of the Dutch TT, partnering Guthrie who retired while leading the race. The finishing order was Gall (BMW) in first place followed by Daniell after a most consistent ride. Two weeks later in the 350 class of the German Grand Prix Harold scored his first works victory ahead of team mate John White. Any satisfaction was later soured by Guthrie's tragic death in the Senior event.

For the 1938 season the Norton team comprised Harold Daniell, Freddie Frith and John White. The work's machines had undergone major changes since 1937, the obvious external difference was telescopic forks to replace the traditional girders. Extensive testing was carried out at the Donington Park circuit and Harold's teammates were generally enthusiastic about the arrangement despite the lack of damping and limited movement. However Harold held strong reservations about the changes and felt more at ease using a Lancefield-prepared practice hack at the TT. He recorded some impressive times with it but during a practice session he crashed in spectacular fashion. The bike was destroyed and Harold was more knocked about than he dared to admit.

Monday's Junior TT saw an emphatic Velocette win by Stanley Woods which was their first Junior victory since Freddie Hicks' win back in 1929. Hicks was a talented engineer and a competent rider and it was something of a surprise when he moved to AJS in 1930. He had scant success with them in his first season, and the following year he was tragically killed in a crash at Union Mills during the Junior TT.

The 1938 Junior TT gave Woods his ninth TT win, backed up by teammate Ted Mellors. Meanwhile Friday's Senior race was shaping up to be one of the most exciting in years. Norton's most serious challenge was expected to come from Georg Meier on a work's BMW and Velocette-mounted Stanley Woods. Meier was in the process of changing a spark plug on the BMW at the warm-up period when he stripped the thread of the cylinder head. Retirement was the only option which saw Meier become a spectator from the grandstand. BMW's 1938 foray had proved to be disastrous when during practice Meier's teammate Karl Gall crashed heavily and fractured his skull. He returned to the Island for the 1939 TT series, only to crash again in practice, this time fatally.

Previous year's winner Frith set a cracking pace from the outset and until lap five he held a narrow lead over Woods on the Velocette. On lap six a relentless Woods assumed first place and seemed poised for a Velocette victory. It was not to be. Daniell had not featured strongly in the opening stages and on the seventh circuit he came through with a record lap of 91 mph to beat the Irishman by sixty seconds. This was a remarkable win for Harold when one considers his practice crash and lack of time to familiarize himself with his actual race bike.

In the Junior TT of the following year the situations were reversed. Despite Daniell recording the fastest lap at 85 mph, it was Woods who came through to score his tenth TT victory. The winning margin was a mere 18 seconds after three hours and ten minutes of inspired riding. This was Stanley Woods' final year of road racing, while for the remarkable Harold Daniell there were still triumphant years as a Norton teamster ahead of him.

The outbreak of war in September 1939 put an end to any racing activities, much to the disappointment of many aspiring riders. Harold was keen to enlist in the armed forces then to his dismay he was rejected on the grounds of defective eyesight which seems rather ironic that the current holder of the lap record on the Isle of Man was deemed unfit for military service. Eight long years were to elapse before Harold resumed an interrupted career where once again he was in the top echelon of road racing.

Johnny Duncan (1935)

Johnny Duncan

1933 Junior TT

The 'quiet man' of the Norton team, Scotsman Johnny Duncan spent only one season with them as a works rider. While other Norton teamsters received a wealth of publicity, in Duncan's case very little information is available. He was born in 1904 at Newmacher, a provincial town near Aberdeen. His racing forays began in 1926 with a 350 cc Cotton.

Duncan's career was unique in that from 1932 he rode exclusively on 'works' machines for four manufacturers; Raleigh, Cotton, New Imperial and eventually Norton. Raleigh, Cotton and New Imperial were examples of manufacturers who struggled through the Depression years and did not re-appear after World War 2.

The Isle of Man was Duncan's main focus, as well as some continental events. His best Raleigh placing was eighth in the 1928 Senior TT. In 1932 and 1933 he was entered by Cotton Motors and in both Senior races he finished seventh. In 1934 he switched to a different make, New Imperial this too was a works entry. Duncan started in all three categories in the TT; the Lightweight, Junior and Senior. With a retirement in every event the outcome could only be regarded as disastrous and despite a luckless year in 1934 he did enough to attract Joe Craig's attention.

The following year saw him as a team member alongside the great Jimmie Guthrie and fair-haired Irishman Walter Rusk. John White was another recruit but found he was restricted by Craig to ride only in selected races. At last Johnny Duncan was blessed with reliable and fast machinery and justified his inclusion with a fourth place in the Junior TT and seventh in the Senior.

Later in the season he took a fine second place in the 1935 European Grand Prix. The venue that year was the demanding Clady circuit at Ulster in Northern Ireland. The Ulster was Johnny Duncan's swan song. Compared to his illustrious teammates he remains virtually unknown outside his native Scotland. Perhaps he preferred to maintain that low profile and turn his back on a motorcycling career. He went back to Aberdeen and set up in business repairing motorcars and lorries, rather than motorcycles. Aberdeen's favourite son died in 1962, aged 58.

Freddie Frith (1936-1939)

Freddie Frith

1939 Junior TT

Lincolnshire stonemason Freddie Frith was born in Grimsby in 1910. His racing career followed a typical pattern for young men with racing aspirations; becoming involved in trials and club events during the late 1920s. He was usually Velocette-mounted and in 1930 he made his Isle of Man debut, which was a portent of a brilliant career when he rode a dealer-sponsored KTT into third place in the Junior Manx Grand Prix. In 1932 he switched to a Norton and gained fifth place in the Senior Manx. This was a commendable effort, for he was riding a 350cc machine in a 500 cc race. His perseverance in the Manx series culminated in 1935 with a win in the Junior event and second place in the Senior behind J.K. Swanston who was a consistent Manx competitor and also a medical practitioner.

Not surprisingly Freddie Frith was recruited into the Norton works team of 1936 along-side Jimmie Guthrie and John White. Frith justified his inclusion in the team with a win and fastest lap in the Junior TT and in a hard-fought Senior he finished third behind winner Jimmie Guthrie and runner-up Stanley Woods (Velocette). Over the ensuing two years Guthrie, Frith and White featured strongly in the continental Grands Prix where the finishing order was usually Guthrie first and Frith second. It was an opportunity beyond compare for the Grimsby lad to serve an apprenticeship in the slipstream of the great Jimmy Guthrie.

Although it was Frith who scored a decisive win in the 500 class of the Ulster GP. and with a record lap of 95mph.

The 1937 TT series provided the usual surprises and excitement with Guthrie heading a Norton 1-2-3 in the Junior event while Friday's Senior was packed with drama. Guthrie assumed a comfortable lead until lap five when he was forced to retire at the Cutting which later became the site of the Guthrie Memorial. Woods was then holding second place and with Guthrie's retirement he inherited the lead. Try as he might he was unable to hold off a challenge by a determined Frith. His final lap was a record-breaking 90 mph which was sufficient to beat Woods by a mere 15 seconds after three hours of incredibly close racing.

Following the TT the Norton team made its regular foray to contest the continental events. In August of that year the German Grand Prix was held on the fast Sachsenring circuit. During the 500 cc race Guthrie held off a strong BMW challenge and was poised to receive the chequered flag when he crashed fatally on the final bend. This was a shattering blow for the team and as a mark of respect it was withdrawn from the forthcoming Ulster Grand Prix.

During 1938 and 1939 it was mainly the Norton team which upheld British prestige in the 500cc class. By then BMW had achieved the reliability to win Grand Prix events. Georg Meier became European Champion for 1938. In 1939 he became the first foreign rider to win a Senior TT on a foreign machine. In second place was an Englishman, the burly Jock West also on a BMW and in a race that he described as his toughest ever was third-placed Freddie Frith. The 1939 Ulster TT was a resounding success for Serafini on the supercharged Gilera four that was miles an hour faster than the opposition. His greatest challenge came from a gallant Freddie Frith who brought his Norton home in second place. The Ulster was the final road race in a fabulous decade and apart from an entry at a post-war Shelsley Walsh hill-climb it was Frith's last race appearance on a works Norton.

Following army service as a riding instructor he returned to racing in 1947. Fred was entered on a potential TT winner with his entry on a 500 Moto Guzzi. A practice crash sidelined him from the TT and other classics. He made a return to racing at the 1947 Ulster Grand Prix and in his efforts to hold off a strong Norton challenge his Velocette suffered a broken valve. On works-supported Velocettes he scored victories in the 1948 Junior TT and in the memorable rain-soaked Ulster TT. In 1949 he gained another victory in the Junior TT riding the new twin camshaft Velocette. This was also the first year of the official World Championships, which resulted in significant British awards. Les Graham was crowned 500cc champion for AJS, while Eric Oliver and Denis Jenkinson became inaugural sidecar champions for Norton.

With victories at the TT and in every classic 350cc race Frith and Velocette were undisputed winners in that category. At the end of the season he announced his retirement and the following year he was awarded an OBE for services to motor racing. The modest Grimsby man opened a successful motorcycle business in that area. Frith maintained his interest in the racing scene particularly the Isle of Man. A measure of his reputation was Joe Craig's offer of a post-race gallop on Reg. Armstrong's 1952 Senior winner. Fred's comparisons between the current works model and the pre-war machines were pertinent. He considered their power output to be markedly similar but the road holding and braking of the new model completely overshadowed its plunger-framed ancestor. Craig's offer was repeated in 1954 when Fred took Ray Amm's Senior winner for a blast along the Mountain Mile. He demonstrated to his old team boss and to motorcycling journalists that the passing years had not dimmed his abilities or his enthusiasm. Freddie Frith died in May 1988 following a brief illness, a respected member of that elite band of inaugural world champions.

Jimmie Guthrie (1931-1937)

Jimmie Guthrie

1934 Senior TT

Few riders have earned greater respect than Scotland's Jimmie Guthrie. During a career that was cut tragically short he gained a reputation of skill, gentlemanly conduct, and astonishing bravery. Even in non-motorcycling households his name was known, not only in Britain but also throughout Europe.

He was born at Hawick in 1897 and from an early age became involved with motorcycling. Following World War One, when he served as a dispatch rider Jim and brother Archie set up as motor engineers in Hawick. Jim's first essay at the TT in 1923 proved unprofitable when his Matchless expired on the first lap.

In 1927 he tried again, this time on a make that never gained TT glory, the New Hudson. The 350 model failed while on the 500 he finished in a creditable second place behind Alec Bennett's Norton. From 1927 until 1937 he rode in every TT, with Nortons in 1928 and 1929 and AJS in 1930. The first of his six TT victories was gained in the 1930 Lightweight TT on an AJS.

In 1931 Jimmie began his loyal association with Nortons when he joined the works team of Woods, Simpson and Hunt. With such an amalgamation of talent it was a rare occasion for a Norton to be beaten. Guthrie made it a Junior/Senior double in 1934, the Junior in 1935, the Senior in 1936 and the Junior in 1937. In total he recorded 22 classic victories and among pre-war riders, only Stanley Woods exceeded this. Guthrie was twice crowned European Champion, today's equivalent of the World Championship. During 1936 and 1937 European manufacturers notably BMW and DKW were challenging Norton's supremacy.

No rider during that period did more to withstand that challenge than Jimmie Guthrie. A portent of German efficiency was forcibly displayed at the 1936 Swedish Grand Prix where Guthrie finished third behind two BMWs. Nevertheless it was Guthrie, Frith and White who were still winning the bulk of the classic races. Following the 1937 TT the Norton team made its way to the continent to contest the classic Grands Prix. The 500 cc class of the Dutch TT resulted in a BMW victory. Guthrie made a gallant attempt to keep ahead but after10 laps his Norton expired with lack of compression.

The next event was the Swiss Grand Prix held on the twisty Bremgarten circuit which placed a premium on handling rather than outright speed. Following their poor showing at the Dutch TT Nortons were vindicated with a Guthrie/ Frith 1-2 in the 500 event. The team journeyed next to Chemnitz to contest the German Grand Prix, held on the fast Sachsenring circuit where Harold Daniell elevated Norton morale with a win in the 350cc race. In the 500 event Guthrie assumed the lead after a battle with Ley on the leading BMW. Ley retired shortly afterwards, leaving Guthrie to consolidate his position. He pitted for fuel at half distance and lost his place temporarily to Gall on a second BMW. Gall experienced problems at that stage leaving Guthrie unchallenged with two minutes in hand.

The vast crowd at the finish line stood up preparatory to applauding the winner but there was no sign of Guthrie. Eventually Gall was flagged in as the winner and there was hardly a cheer for him. When the news came in it was all-bad. Guthrie had crashed, was seriously injured and on his way to hospital. The ambulance journey was a tragic anti-climax. With thousands of spectators clogging the roads it took the ambulance two hours to reach the hospital and sadly Guthrie succumbed to his injuries shortly afterwards.

Various theories have been advanced as to the cause of Guthrie's accident. One was that the rear spindle broke, dislodging the back wheel. Another was that the con-rod failed because the engine had run low on oil, locking it solid and jamming the back wheel. In later years Stanley Woods came forward with his version.

I had coasted to a stop with fuel problems at the bend where Guthrie crashed and witnessed the chain of events .I am adamant that Guthrie was fouled by the German rider Mansfield on a DKW. Guthrie was about to lap Mansfield when the German deliberately moved across, forcing Guthrie into the right-hand gutter. The Norton tangled with a row of saplings, pitching Guthrie off with fatal results.

Woods was first on the scene and found his old teammate and best friend completely shattered. He accompanied Guthrie in the ambulance and on the way to hospital Guthrie lapsed into a coma from which he never emerged. Jim Guthrie's death was particularly tragic as he had made the decision to retire from racing at the conclusion of the 1937 season.

The brilliant Scot's feats are preserved with two significant memorials. One is a bronze statue mounted on a stone pedestal; displayed in his hometown by the citizens of Hawick. The other is a stone cairn with its symbolic broken column erected at the Cutting where Guthrie retired in his final TT race the 1937 Senior. The Guthrie Memorial is an evocative site where enthusiasts can pay homage to one of the greatest on their Isle of Man pilgrimage.

In Germany too, where he was held in great esteem Jimmie Guthrie was not forgotten. Each year floral tributes are laid around the plaque at the crash site in Saxony

The Guthrie name was revived in 1967, the year that Jim Guthrie's son Jimmy won the Senior class of the Manx Grand Prix. Guthrie Junior rarely ventured beyond the Manx races but he demonstrated his ability on the demanding Isle of Man circuit, the scene of so many of his father's triumphs.

Walter Handley (1934)

Walter Handley

1934 Junior TT

In that exciting decade of the 'roaring twenties' no rider attracted more public adulation and media attention than Walter Handley. His brilliant riding and utter determination to finish, regardless of the conditions were legendary. Yet at times he was prepared to scuttle his chances with deliberate disregard for his machine. In a career that spanned 12 seasons he held no great loyalty for a particular make and his list of entrants embraced most leading British manufacturers, plus several European firms. Stories abound concerning his negative attitude to team discipline, coupled with a fiery temper and an attitude at times that was most charitably described as 'awkward.'

Handley was born in Birmingham in 1903 and despite being grossly under-age during World War 1 he attempted to enlist in the army. After the war he found employment at the OK Motorcycle works as a mechanic and road tester.

He was a member of the OK team that was entered for the 1922 Lightweight TT. Walter raised the lap record by 5mph to 51mph only to retire with a broken valve. At the end of the season he scored a resounding win in the 250 class of the Ulster TT. Following a difference of opinion with his employer Walter switched to the Coventry-based Rex Acme firm. It proved to be a rewarding experience for Handley who went on to score wins in the 1925 Ultra Lightweight, the 1925 Junior and 1927 Lightweight TTs. His second place in the 1926 Senior was a classic example of his utter determination to finish. His Rex Acme V-twin oiled a plug at the start and due to its awkward placement he lost seven minutes changing it. Then in a ride that defied belief Walter took the evil-handling Rex from 22nd place to a meritorious second behind Stanley Woods' Norton.

In 1929 he signed with the Swiss Motosacoche firm with options to ride other makes. On a 350 AJS he rode brilliantly to finish second to Freddie Hicks' Velocette in the Junior TT. For the 1930 TT he was entered on a Belgian FN and with its non-arrival in the Isle of Man it appeared he would be a non-starter. Fate then took a hand through the generosity of dealer/entrant Jim Whalley. He was entered on a works Rudge, partnering their team of Graham Walker, Ernie Nott, Tyrell Smith and sportingly he gave up his own ride in favour of Handley.

Earlier in the week the Rudge team scored a comprehensive 1-2-3 in the Junior event and was looking confidently towards Friday's Senior race, which began in fine but deteriorating weather that degenerated into torrential rain. Walter barely eased his pace and led through-out to win the Senior TT and make fastest lap. In 1932 Rudge-mounted Handley finished second in the Junior TT behind Stanley Woods' Norton. Apart from a popular win for Jimmie Simpson in the 1934 Lightweight that was the final taste of victory for Rudge Whitworth in the Island.

Handley then switched his allegiance to Velocette. Their 350 KTT had scored decisive wins in the 1926, 1928 and 1929 Junior TTs. Private owners were thus able to purchase an 'over the counter' racer whose specification compared closely to the current works machinery. Apart from a win in the 350 class at the Ulster Grand Prix Handley spent a fairly lean year with Velocette.

Norton's problems at the end of 1933 were about riders rather than machinery. Tim Hunt had suffered injuries at the Swedish Grand Prix that ended his career completely. Stanley Woods was disenchanted with team orders that dictated who was to win and as a result he signed with the Swedish Husqvarna team.

To bolster the Norton team Handley was signed up as a member for 1934. Their first outing was an Irish event, the Leinster 200 and although Handley was to make his debut there he failed to appear. Next outing was the North West 200 and this time Handley started in the 500cc race, holding second place until clutch problems resulted in his retirement. The team then moved to the Isle of Man.

This was an auspicious series for Guthrie who scored a Junior/Senior double and a disaster for Handley who crashed at Governor's Bridge on the sixth lap of the Junior. He severely damaged his nose which was enough to put him out of Friday's Senior. He recovered sufficiently to start in the Belgian Grand Prix and in other European classics. His final Norton appearance was at the Ulster Grand Prix where he finished third behind Walter Rusk on a works Velocette and veteran Charlie Dodson, making a rare Norton outing. The Senior Ulster proved to Handley's final Grand Prix finish.

For the 1935 season he returned to Velocette and was looking forward to better fortunes with the Hall Green concern. Alas it was not to be. Following a TT practice session he leant down to adjust his rear brake whilst on the move. His fingers became trapped in the chain and were badly mangled, enough to side line him from another TT.

Handley's motorcycle career appeared to be waning. He directed his energies to motor racing and flying although in 1937 he was coaxed out of retirement to put a modified 'Empire Star' BSA through its paces at Brooklands. It was a surprising move for BSA who had shunned any race appearances since their 1922 TT debacle. Walter dominated his event and in the process he earned a Gold Star for lapping Brooklands at 100 mph. And thus was born the evocative 'Gold Star', a machine that achieved total dominance in post war clubman racing. This Brooklands venture proved to be Handley's final race appearance, in a later event at the same meeting he collided with another rider and suffered painful injuries in the ensuing crash.

At the outbreak of war Handley was commissioned in the Air Transport Auxiliary. The ATA was an organization of male and female pilots who performed vital duties in ferrying new and repaired aircraft from factory and maintenance units to operational squadrons.

In 1941 Flight Captain Handley was at the controls of a Bell P39 Airacobra, an unorthodox single-seat fighter with its engine mounted behind the pilot. It was reported that on take-off, Handley climbed too steeply, stalled and with insufficient height to recover the aircraft crashed and burned. The great Walter Handley died as spectacularly as he lived.

Visitors to the Isle of Man can view a commemorative seat on Bray Hill erected to his memory. The celebrated motorcycling journalist and TT winner Graham Walker dedicated it shortly after the war and its gold plaque ends simply with the words: *None passed this way more bravely.*

Tim Hunt (1931-1933)

Tim Hunt

1931 Junior TT

Tim Hunt burst on the racing scene in the style of a shooting star. His star burned brightly for a few seasons and then abruptly it was extinguished. He was born in Manchester in 1908 and christened Percy but from an early age he was labelled 'Tim' after a comic book character named 'Tiger Tim'. Apart from his riding ability he was also blessed with wealthy parents, to the extent that Mrs. Hunt was quite happy to provide the funds to launch Tim on his chosen career.

His first road racing success was the Senior class of the 1927 Amateur TT (as the Manx Grand Prix was then called). In 1928 mounted on a new CS1 Norton he won again and in the process he broke Stanley Woods' Senior TT lap record. These efforts did not go unnoticed by Norton talent scouts and Tim was eagerly signed up as a Norton teamster. A measure of Hunt's versatility and also the CS1 Norton's was demonstrated when he rode his race bike in the 1928 Scottish Six Day's Trial and collected a first class award in that strenuous event.

Despite a lack of engine skills it was rare for him to suffer mechanical failures; his great ability was in his natural riding talents. In 1931 he became the first rider to score a Junior/Senior double at the TT, with fastest lap in the Junior. The Senior was also memorable for teammate

Jimmie Simpson who recorded the first-ever 80mph lap. Not surprisingly he failed to finish, once again reinforcing his nickname 'Unlucky Jim'.

In the continental races it was Hunt and Woods who generally rode the 500s while Guthrie and Simpson handled the 350s. Such was Norton's domination during the era that it was headline news when they occasionally finished out of the money. Of the twelve classic races of 1932 that Norton contested they won eleven. The team for the 1933 season remained unchanged. Stanley Woods was once more the star of the TT series after repeating his 1932 Junior/Senior double. From there the team moved to the continent to contest the classic events. The spoils were fairly evenly shared between Hunt and Woods in the opening rounds with Hunt having success in the 350 Swiss Grand Prix and 500 class at the Dieppe and Belgian Grands Prix.

Team orders for that year's Ulster Grand Prix were for Hunt to win the 500 race. Irishman Woods was furious with the arrangement, particularly in front of a home crowd. Ignoring team orders he went into an early lead and in the process lapped at a record 89 mph. Hunt eventually caught Woods and for several laps they raced neck and neck. Eventually Hunt was forced to retire with a broken steering damper, leaving Woods free to win with a clear conscience.

From Ireland the team travelled to Saxtorp to contest the Swedish Grand Prix, which was also that year's European Grand Prix. It proved to be a disaster for Nortons and particularly so for Tim Hunt. Their greatest challenge was expected to come from the Swedish Husqvarna team of Sunnqvist and Kalen. As the race progressed, Woods, Hunt and Sunnqvist were slip-streaming one another down the main straight when a rider they were lapping slowed dramatically. Woods and Sunnqvist both made a miraculous avoidance but Hunt careered into him. When Tim was retrieved by the ambulance team he was found to have suffered a seriously smashed hip while the other rider, a Norwegian was tragically killed.

This unfortunate incident was a sad ending to Hunt's career. He spent three months in a Swedish hospital before being transferred by plane to England for further surgery. Medical technology was not advanced enough in those days to plate shattered limbs, resulting in Tim spending the next five years in and out of hospital. After many operations he finished up with one leg shorter than the other. He never raced again. Tim's contemporaries were agreed that if he could have been more serious about his attitude to racing he would have been the greatest ever. Tim believed in living life to the full and being serious was just not in his make-up.

Joe Craig held Tim Hunt in the highest regard and years later he confided to his current champion Geoff Duke that of all the riders he was involved with, he regarded just two as having that innate riding excellence; 'Tim Hunt and Geoff Duke'.

Walter Rusk (1933-1935)

Walter Rusk

1935 Junior TT

For such a small country Ireland produced an abundance of talented riders. One of Ulster's favourite sons was undoubtedly Walter Rusk who is still spoken of warmly to this day. Win or lose he raced for the thrill of it and with his shock of fair hair he was labelled the 'Blonde Bombshell'.

Walter was born in Belfast in 1910 and from an early age he developed a passion for speed, particularly on two wheels. His first road race win was the 1931 Temple 100 on a 500 Norton. His ability caught the attention of two manufacturers, firstly AJS who provided a factory bike for the 1931 Ulster Grand Prix. Mechanical problems resulted in a retirement and for 1932 he was on a 250 works Rudge. The Lightweight event developed into a race-long battle with the works New Imperials of Ted Mellors and Syd Gleave. Walter led the race until half distance when he was slowed with magneto problems. Finally, he finished third after a typically determined ride.

In 1933 he was entered for the Senior TT on a Sunbeam. The 'Beam failed to finish but his dashing style caught the attention of Joe Craig. The outcome was an invitation to join the Norton works team alongside Stanley Woods and Tim Hunt. His first race on a works Norton was the 1933 Ulster. Team orders were for Hunt to win the Senior class which was an arrangement that greatly displeased Woods, who was naturally keen for success in front of a home crowd.

Right from the start Hunt and Woods became involved in a wheel-to-wheel battle which finally ended with Hunt's retirement. Rusk made a strong challenge to pass Woods but the wily Dubliner held his lead to the chequered flag. To be beaten only by Woods it was plain to see that Rusk was a serious contender.

Still on Nortons, Rusk won the 350cc class of the 1934 North West 200. This was an important event in Irish races and was held just prior to the TT in June. Walter was entered on Velocettes for the TT in which he finished seventh in the Junior. On the new 500 Velo he prevented a Norton clean sweep by finishing third in the Senior, and later in the year he scored a memorable win in the Senior Ulster and with a record lap of 92mph.

For 1935 he re-joined the Norton team and at the NW 200 he scored another win in the 350 race. Then it was off to the TT where Nortons were fielding a four-man team of Guthrie, Rusk, Duncan and White. The Junior was a predictable Norton 1-2-3 of Guthrie, Rusk and White and due to impossible weather conditions Friday's Senior race was postponed until the Saturday. It was a major disappointment for hundreds of spectators who were unable to extend their travel and accommodation bookings. For those who could, it proved to be one of the most dramatic of all Senior TTs.

Earlier in the week Stanley Woods scored a convincing Lightweight win on a 250 Moto Guzzi. He was entered for the Senior on the rear-sprung twin Guzzi, which was regarded as fast yet fragile and unlikely to seriously challenge the Nortons. For six of the seven laps Guthrie held a commanding lead. Woods had climbed to second place but appeared well beaten. Woods then called on all his resources to put in a final record lap to beat Guthrie by a margin of four seconds after three hours of intense racing, while two minutes in arrears was third-placed Walter Rusk.

The 1935 season culminated with the Ulster Grand Prix which was also that year's Grand Prix d'Europe. Rusk was naturally anxious to repeat his 1934 Velocette victory and he and Guthrie set a furious pace well ahead of the field. Both came to grief on melting tar at Muckamore; Guthrie was quickly up and away, however Rusk suffered a fractured arm, which proved to be a serious injury and was enough to put him out of racing until the 1938 TT. Despite not being 100% fit he returned to the series with a 250cc OK Supreme. Those OKs were notoriously fragile and apart from Frank Longman's 1928 Lightweight victory were rarely among the winners; factors that resulted in Rusk's retirement in Wednesday's Lightweight TT. He enjoyed better fortune riding Nortons in that year's continental races and for 1939 he was entered for the TT and the Ulster on a works AJS. His mount for the Junior TT was a new Mk.8 Velocette.

The blown AJS V4 had undergone serious changes since its last outing at the 1936 TT., including water cooling. Walter Rusk and Bob Foster each rode one in the 1939 Senior. They were down on speed and at least they held together for the entire race, with Rusk finishing 11th and Foster 13th. At the 1939 Ulster Grand Prix Rusk gave the fearsome AJS a moment of glory when he initially led the 500cc race. In so doing he became the first rider to lap a British circuit at 100 mph. But the pounding the Ajay received was just too much for it. Walter was leading by 20 seconds when he was forced to retire with a broken fork link.

Ulster was the final event in a memorable decade of racing. Days later World War 2 began and seven long years were to pass before those daring young men donned their leathers again. Many of the old faces were missing and sadly Walter Rusk was one of them. He was not prepared to sit out the war at home and volunteered for the RAF. Walter was accepted for the pilot's course and on completion of his primary training he then moved on to advanced training at No.7 Training School at Peterborough. On an early morning flight a Hawker Hart crashed into a hillside, killing its student pilot. Walter Rusk's death robbed Ulster of one of its best-loved racers.

Jimmie Simpson (1929-1934)

Jimmie Simpson (1929-1934)

1932 Senior TT

Maker and breaker of a host of record laps Jimmie Simpson almost invariably led his races from the start. In those races in which he finished he won them by big margins. Sadly Jim's enterprising driving methods were just too demanding for the current machinery. This was especially the case in the Isle of Man where his consistent ill fortune earned him the title of 'Unlucky Jim'.

His inaugural TT was in 1922 when he was entered on a 500 Scott. The Yorkshire based firm were keen supporters of the TT. Their product was a water-cooled twin cylinder 2-stroke; and was a format from which they never wavered until their eventual demise in the 1970s. In 1912 and 1913 a Scott finished first in the Senior TT. By the 1920s, apart from the odd leader board finish they were never to repeat those earlier successes. Simpson's 1922 Scott experience lasted only half a lap when his fuel tank split.

In 1923 he began an association with AJS which continued until 1928 when he switched to Nortons. Jim's first AJS outing in the 1923 Junior TT was typically Simpson; leading the race by over a minute before he retired following a race crash. His sole consolation was a record lap of 59mph. In between his racing activities Jim was employed by AJS as a motorcycle tester. His philosophy, in respect of his occupation was brutally simple; *Bust them and you'll find where their frailties lie; nurse them and you never will.*

Simpson pursued the same attitude in his racing forays. Had he been prepared to show restraint it's a fair assumption he would have gained less record laps and more Tourist Trophies. In a 13-year Island career he rode in 26 TTs and finished first in just one, the 1934 Lightweight. In only two years did he fail to score a place or set up fastest or record laps. He was the first rider to turn race laps at 60, 70 and 80 mph average. The biggest margin where a Simpson record lap beat the race average was in the 1924 Junior TT when Simpson's AJS lapped at 64mph compared to Kenneth Twemlow's winning New Imperial at 55mph. Predictably, Simpson failed to finish.

In 1929 he began a new career with Nortons and Jim's unerring bad luck coincided with a temporary slump in Norton's fortunes. Charlie Dodson gave Sunbeam their last Island victory in the 1929 Senior and in the 1930 TT series Rudge were on the crest of a wave and seemingly invincible. Simpson's third placing prevented a Rudge 1-2-3 in a rain-soaked Senior TT.

However this proved to be Rudge's last Senior victory and by 1931 Nortons had achieved a degree of reliability that completely dominated racing until well into the decade. Importantly for Simpson he was at last riding a machine that not even he could break. Simpson recorded leader board finishes with a third in the 1932 Senior and second place behind Stanley Woods in the 1933 Senior. In 1934 he made the decision to retire at the end of the season. Following a career highlighted by spectacular crashes he realized he was fortunate to be nearing the end of it still in one piece. As it happened 1934 was memorable for Jimmie Simpson, a year in which his personal gremlins finally relented.

Two runner-up finishes behind Jimmie Guthrie in the Junior and Senior TTs were most satisfying. More importantly he finally won that elusive Trophy with a win and fastest lap in the 250 Lightweight, giving Rudge their final Isle of Man victory. A pleasing bonus to round off the season was a string of firsts in the 350 class of the Grands Prix of Holland, Belgium, Switzerland and Ulster. Following his retirement from racing Simpson still maintained a close association with the TT in his position with Shell's competition department.

John White (1935-1939)

John White

1938 Senior TT

By the late 1980s there was just one remaining member of the elite 1930s Norton works team. He was John White the Cambridge graduate who rode Nortons from 1935 to 1939. Born in Radlett Hertfordshire John White became involved with motorcycling from an early age. His inspiration to ride in the Isle of Man came from fellow student Malcolm Muir. Both were members of the Cambridge University Auto Club and following Muir's win in the 1931 Senior Manx Grand Prix White made the decision to ride in the 1932 event.

Somehow he scraped together the forty-five pounds required to purchase a second hand KTT Velocette which was a sizeable sum in those days, although it was not an auspicious debut for the newcomer. In the Junior race he came off on two occasions and eventually retired. He returned the following year with a much-improved KTT and on it he led the Junior Manx and lapped at a record 75mph. Once again he was forced to retire, following a crash on the Mountain climb at the Gooseneck. White rode the 350 Velo in the 500 event and distinguished himself by finishing fifth, even after two crashes during the race. As a result of these aberrations his university chums labelled him 'Crasher' White. This was a somewhat unfair appellation that continued throughout his career.

Nortons were sufficiently impressed with White's rides to provide machines for the 1934 Manx. At the time British factories all signed a bond that they would not support riders in the Manx but as White commented.... *the ink was hardly dry before Nortons broke the agreement.* White had his first Manx success on Nortons with a win in the Junior event. But things did not go smoothly in the Senior. White led the race until he was halted briefly to replace a spark plug lead that had come adrift. Determined to make up the lost time he came off in a big way at Union Mills and considering his speed was in the vicinity of 100 mph he was fortunate to walk away from it.

Nevertheless John White became a member of the 1935 works team. His first outing was the Leinster in Ireland which was treated by the factories as a warm up event before the TT. When his works bike did not arrive in time for practice he was obliged to use his own hack Norton. As a result he began the race on an unfamiliar factory machine and crashed out. The bike was damaged which infuriated Joe Craig who wanted to drop him from the team for the TT.

After a spirited discussion Nortons relented to the extent of providing a bike for the Junior but not the Senior. The Junior was a clean sweep for Norton with a Guthrie, Rusk and White 1-2-3. Despite his Junior success White was not included in the team for the Swiss or German Grands Prix that year. He did get a bike for the Belgian Junior Grand Prix which he won with a record lap at 84 mph. At the time he was a biology master at Lydney Grammar School and was fortunate to have an indulgent headmaster who allowed him time off to compete at the TT and the main continental events.

For 1936 White was again a Norton team member and with no restrictions as to where he could ride. This was the year that Nortons appeared with rear springing on their works machines and riders were all agreed that it contributed significantly to comfort and handling. Freddie Frith scored a convincing win in the Junior TT on the new model and backing him up in second place was John White.

Following the TT, Nortons began the rounds of the classic Grands Prix. White was victorious at the 350 Dutch TT ahead of Frith by less than one second. At the Ulster

Grand Prix Frith scored an emphatic win in the 500 race. In the 350 Ulster which was run concurrently with the Senior White fought a tremendous battle with Ted Mellors on a works Velocette. White crashed on the fourth lap and Mellors retired with chain trouble, leaving the way clear for eventual winner Ernie Thomas on a second works Velo.

For the 1937 season the works Nortons featured twin-camshaft valve operation, an innovation that markedly improved their performance. Guthrie led a Norton 1-2-3 in the Junior TT, with Frith second and White third. Friday's Senior race was memorable for Guthrie's retirement and Frith's heroic last lap victory over Woods on the Velocette, while in third place once again was 'Crasher' White.

White had the satisfaction of winning the 350 Dutch TT for the second time and in the process he set a lap record that stood for 12 years. On the following weekend he won the 350 Belgian Grand Prix, also with a lap record. That year's German Grand Prix was held at the Sachsenring at Chemnitz. Nortons dominated the 350 event with Harold Daniell winning from White by the margin of just one second. Following the 500 race and Guthrie's tragic accident as a mark of respect the Norton team did not contest that year's Ulster Grand Prix.

Prior to the 1938 TT Nortons carried out extensive testing of a new telescopic front fork to replace the current girder pattern. Frith and White expressed their preference for the new forks although Daniell was not totally convinced. Despite his reservations he won the Senior TT with a record lap at 91 mph which stood unbroken until 1950. In third place in both Junior and Senior events was the reliable Freddie Frith ahead of the consistent White.

In the two continental races that White contested, the 350 Dutch and Belgian he scored wins in both, although the Ulster was something of a disaster for Nortons. White crashed in the 350 race which Mellors won, amassing sufficient points to become 350 European Champion. Frith's Senior race was also disappointing. He put in a stupendous lap at 98 mph but crashed when a sudden shower on the circuit caught him out, leaving the way clear for Jock West to claim victory for BMW.

With the prospect of war looming, Nortons pulled out of racing for 1939. At the last minute they relented and provided last year's TT machines for White, Daniell and Frith. White had also arranged to ride a blown NSU twin in the Junior TT. The ungainly rigid framed NSU never impressed during practice and expired in its race but the Norton brought him home in fifth place in the Senior.

White scored his last race win at the 1939 French Grand Prix, which was held at Rheims and was not a fully supported event. Shortly afterwards war was declared, and despite being in a reserved occupation White was anxious to be involved in the war effort. Eventually he was released from school teaching and joined the ranks of experienced motorcycle instructors. Later he transferred to the engineers and finished the war as a captain.

He never raced again. By that time he had married and did not consider racing to be a suitable occupation for a married man. John White's TT record with Nortons was a model of reliability with eight finishes from eight starts, the majority of them on the leader board. His sole TT retirement was the 1939 Junior when his blown NSU expired early in the race. He remembered his teammates with affection, particularly Guthrie and Frith but with reservations about Daniell. His only regret was that he would have preferred to ride for Velocette. In his opinion they would not have restricted his race appearances the way Norton did. They may have even paid him a salary which Norton never did.

Stanley Woods (1926-1933)

Stanley Woods

1927 Senior TT

In a racing career that began in 1922 and ended in 1939 Ireland's Stanley Woods established himself as possibly the greatest of his era. His record of 10 TT wins surpassed that of Jimmie Guthrie, his nearest rival with 6. In continental events Woods was equally dominant. Undoubtedly his most profitable period was spent as a Norton teamster, a time that yielded five Isle of Man wins plus a host of victories on the continent.

Stanley Woods' Isle of Man career began in 1922 and in unique circumstances. He overcame the lack of a bike by sending letters to various makers, offering his services as a potential works rider. Not surprisingly the unknown teenager was not besieged with offers, with one exception. F.W. Cotton, founder of the Cotton Motor Company did provide a mount for the Junior TT. Woods began the race in determined fashion and following a series of mishaps that would have dismayed most individuals the tyro Woods did finish the race.

Cottons were sufficiently impressed with his efforts to offer him full works support for the 1923 Junior TT. Despite running out of road and bending the bike's forks during the race, Woods came through to win his first TT. The next two years were generally unrewarding for Woods. During that period he rode in trials and road races on Royal Enfields and New Imperials respectively.

His fortunes improved markedly in 1926 when he was offered full-time employment with Nortons. Despite the near-obsolescence of their current works bikes Woods gained his first Senior Trophy. He was unfortunate not to repeat his success the following year which was the debut of the CS1 camshaft Norton. Woods set off in meteoric style to lead second-placed Alec Bennett by over 3 minutes. He refuelled on the fourth lap but his pit crew simply advised him to maintain his present pace. Unaware of his actual lead he pressed on, only to experience clutch trouble and retire on the fifth lap.

Embittered by the outcome it at least opened Woods' eyes to the pitfalls of haphazard signalling in road races. The Senior TT debacle set off the train of thought that led to the evolution of his unique signalling system.

Alec Bennett took the CS1 to its only Isle of Man success and the period between 1927 and 1930 saw the dominance of firstly Sunbeam and later Rudge. That all changed dramatically with Tim Hunt's Junior/Senior TT double in 1931. Stanley Woods emulated Hunt's feat in 1932 and to reinforce his record he repeated the performance in 1933. These successes would normally have been financially rewarding but the Great Depression was biting deeply into all aspects of commerce. Trade bonuses and prize money had been reduced considerably and for a professional rider it was absolutely vital to seek out the best possible retainer. At the end of the 1933 season Woods made the bold decision to leave Nortons. Despite his run of success with them he considered he was worth far more than they were prepared to offer. Another situation that irritated him was the stricture of riding to team orders.

Other manufacturers, notably Husqvarna and Moto Guzzi were keen to obtain his services. For 1934 he signed with Husqvarna for the TT and other continental events. Woods was impressed with the performance of the V- twin Husqvarna which had achieved a degree of success during the1933 season. His TT misfortunes began even before the 'Huskies' left Sweden when the lorry containing the works bikes dropped from its crane sling during loading on to a steamer. They were extensively damaged and as a result their carburetion and other details were not fully sorted until race eve. Despite this handicap Woods made fastest lap in a wet Senior race only to run out of fuel on the final lap.

Woods' Husqvarna experience proved to be generally unrewarding although with Moto Guzzi in 1935 it was quite the reverse. In Wednesday's 250 Lightweight event Woods scored the first of Moto Guzzi's 10 Island victories. Conditions for the Lightweight were dismal enough but for Friday's Senior they were so atrocious that the start was postponed until Saturday. As expected, Jimmie Guthrie built up an early lead over Woods and at the end of lap six Guthrie's lead was 26 seconds. It was assumed that Woods must stop for a second time to refuel at that stage which would place him even further behind Guthrie. As a result Guthrie was given the signal to ease up on his final lap. There was immediate drama when Woods sped past his pit without stopping. Riding as never before he turned a 26 second arrears into a four second advantage and was the margin by which he won the 1935 Senior TT, surely the most dramatic TT of all time.

Following his success in the 250 and 500 categories, Woods was keen to add a competitive 350 to his entries. On two occasions he made overtures to Velocette to sample their bikes and to his surprise his requests were politely declined. Walter Handley was their No.1 rider at the time and was a non-starter at the 1935 TT after inflicting damage to his hand. Handley's misfortune became Woods' opportunity. Velocette managing director Percy Goodman then issued an invitation for Stanley to make an appraisal of their machines. It was also revealed that Handley was instrumental in blocking Woods' earlier approaches to the firm.

Woods was an astute judge of racing machinery and his assessment of the Velocette was brutally honest. It was a fine motor he declared but the gear ratios were unsuitable and the handling was unacceptable. Velocette were prepared to accept his advice as to any modifications and during the winter of 1935-36 these were put in place. The results were revealed on the works bikes at the start of the 1936 season. They featured a twin-camshaft 350 and a completely new single-cam 500 but importantly the works machines featured swinging arm rear suspension with oleomatic dampers. Their design set the pattern that continues to this day.

Woods was quietly confident that the twin-cam 350 would be unbeatable on the day. It was not to be. Halfway through the first lap he was forced to retire. Teammate Ted Mellors took up the challenge to the work's Nortons and finished a creditable third. Woods' Senior race was more profitable. Atileast when it developed into a traditional Woods/Guthrie contest and at the finish Guthrie had reversed last year's result, although Woods again made fastest lap. As a point of interest, Woods made fastest lap in three consecutive Senior TTs (1934-36) and on three different marques; Husqvarna, Moto Guzzi and Velocette. For three consecutive years (1936-38) Woods battled mightily to win a Senior Trophy for Velocette and each time he finished a gallant second. On the 350 he had Norton's measure and scored wins in the 1938 and 1939 Junior TTs to bring his Isle of Man victory total to ten. An important aspect of the 1939 TT was Stanley's opportunity to put in a solitary practice lap on the brand-new Velocette 'Roarer'. He was impressed with its performance, in particular the handling but the blown twin still needed further development to make it competitive. He chose to ride the proven single in the Senior and finished in fourth place, in what was to be his last finish at the TT.

Ulster was generally a happy hunting ground for the canny Dubliner.and it was fitting that his chequered career should finish there with a masterly win in the 350 class of the 1939 Ulster Grand Prix. A few days later war was declared and like so many riders his career suffered postponement. He came tantalizingly close to making a comeback at the 1947 TT. Woods had entered Freddie Frith on a 500 Moto Guzzi, similar to the model on which Stanley had won the 1935 Senior.

A practice crash at Ballacraine sidelined Frith from the TT and for most of the season. Significantly it left the Guzzi without a rider for the TT. Stanley agonized over a sleepless night whether he should take over Frith's entry, knowing that he stood a good chance of winning despite the enforced lay-off due to the war. Eventually he decided against it but he maintained his association with Moto Guzzi and frequently rode their bikes at Italian test tracks and in the Isle of Man.

Even in his 70s the grand old timer still made parade appearances on "his" 500 Velocette. The historic bike is the actual 1937 ex-works model and forms part of a fabulous collection of Velocette racers, including the 1939 'Roarer' owned and meticulously restored by Velocette devotees, Ivan Rhodes of Derby and his sons, Grahame and Adrian.

Right up to the time of his death in 1989 he retained that enthusiasm, enhanced by an infallible memory and razor-sharp faculties. Stanley Woods was remembered with affection by generations of enthusiasts.

That Lap at 91mph

Harold Daniell, TT Record Holder tells how it was done

by
H. L. Daniell

Quite a number of people have expressed to me some surprise that my record circuit for the T.T. Course was made on the last of seven laps. There seems to be a feeling that it would be more natural to go faster at the commencement of the race than at the end, the idea being, I suppose that both man and machine should then be in better fettle.

My answer to that is covered in several points. First, my own personal inclination is to start steadily and warm up as I go along. I like to reserve my mechanical and mental resources in the earlier stages, keeping something in hand to deal with the need for any special effort that may arise as the race nears its conclusion and the situation, as presented by the Leader Board is becoming clearer.

Secondly there is the obvious fact that, providing you aren't tiring and that the machine is still going as well as it did at the start, it is easier to do the last lap quicker than any other.....for these reasons. Lap 1, being from a standing start is a dead loss as regards records. Lap 2 is a flying one but the petrol tank still carries a heavy load of fuel. Lap 3 entails a stop for refuelling. Lap 4 is from a standing start. Laps 5, 6 and 7 are all flying ones but the load in the tank is getting progressively lighter. In short, between the refill and the finish some 60-odd lbs. of high-up, shifting weight have been shed.

Another point is that, as the race gets older the number of competitors dwindles and the road becomes clearer. Again, the pit equipe is able to give you a better idea of how you stand and if they call for more speed in the closing stages it is easier to respond if you haven't already exhausted yourself by piling on sail right from the start.

A further factor is that one can regard the opening laps as just a bit more practise. When you are taking part in two races and have to divide the training period between Junior and Senior mounts there isn't any too much time to get used to either. What with bad weather, hordes of other riders on the course, experiments with jet sizes, plug "chop-outs" and so forth it may well be that one may never find out what the Senior model will really do over one lap until the race is actually under way. Even then the conditions are quite different from morning or evening practising.

Why this is so I am not quite certain but I do know and many other riders agree with me, that the bike always seems to go slower on race days than it did in practising. I think it must be something to do with the light -- on the principal that that one always seems to be travelling faster at night. This theory is supported by the fact that this phenomenon is more pronounced in the Manx Grand Prix races when the shorter September days make the contrast between early morning, midday and evening conditions even sharper.

At any rate it always takes me some time to get the "feel" of the model and accustom myself to the race day atmosphere; then, one's confidence in the machinery and in oneself increases, so does the urge to go quicker become intensified. But friends who watched the 1938 Senior Race have often told me that I did not appear to them to go round the corners any faster on the last lap than on the first. My reply is that their observation was probably quite correct...I didn't! And that is where the difference between a short circuit event and a long distance race like the TT.

In the former you have a small number of corners that you are repeatedly negotiating and to get anywhere at all you have to do your best to scramble round them as fast as you can. But that type of racing isn't likely to pay off on the I.O.M. circuit. Over there seconds are saved, not on the bends themselves but *between them!*

In recent years a great deal has been said and written about taking the right "line" through a corner. Many young riders appear to think that each curve on the course has an invisible tape stretched along the road surface and all that is required is to follow it. I do not dispute that there is a best way through every bend and I would never discourage anyone from seeking it but my advice is to concentrate more on the approach to the corner and the manner in which you come out of it.

Take, for instance, Hillberry as it is now. It *has* been ridden "flat" in top gear when one would certainly save time on the approach and on the corner itself, but that advantage would be lost in extricating oneself from the bank on the far side and finding oneself completely off course for the left-handed climb to Cronk-ny-Mona, which in any case calls for a drop in gear. Much better to engage third before the corner and come out accelerating along the correct approach to the hill beyond.

That is my policy everywhere. Whenever I approach a corner that offers a choice of ratio, I select the lower one and drive the machine through the bend rather than allow the bike to drive me - perhaps in a direction I don't want to go! This method may seem harder on the engine.... but if the engine won't take it, then you won't win the race anyway!

Now let us embark on a flying lap and see how this precept works out. We will assume it is the last lap of a Senior Race. There is a minimum "upstairs" in the tank, it's a fine day and you still feel fit and fresh.

As you come down the Glencrutchery road, building up to peak revs in third (115 m.p.h.) you spot your pit signal which is calling for an all-out effort. When the grandstands have slipped astern you stub the lever into "high" and prepare for Bray Hill. The bumps by the St.Ninian's crossing will send you flying – so remember your face. True you've got a chin pad but there are all sorts of furniture around the handlebars that will try to knock out your teeth if you are not careful.

In 1938 the way down Bray Hill was in the middle of the road, moving over to the left approaching the foot to swing right across the cross roads. Nowadays it is possible to keep left all the way down which makes the navigation of the bottom less tricky.

The descent to Quarter Bridge calls for care; you are dropping through the gears and braking hard. The approach is overhung with trees, the surface is often damp and the light is treacherous.. No fireworks on the bridge --- it is one of the worst corners of the lot. Take full advantage of the camber but look out for wet tar in warm weather, especially on the last lap.

On the straight approach to Braddan it is a toss-up whether to get into top or hang on to third. When we ran on 50/50 mixture it paid to use the fourth cog. On 'Pool" I am not so sure but perhaps it is still best to use top so as to get the full braking effect for the "S" bend over the railway bridge, which is a second gear proposition. The twisty stretch which runs through Snugborough is fast third gear stuff with a short spell in top just before Union Mills which is taken in second, changing up as you pass the Post Office on the way out. The right hander at the top of the rise before Crosby is OK for full blast on top, but it calls for respect and is the sort of bend that demands that you give way if there is anyone just ahead of you.

The "Highlander" I never really found difficult. It is full bore, slap in the middle of the road and take the bumps head-on. Hall Caine's Corner or Greeba Castle, I take in third; it is one of those places that is a lot quicker than it looks. "Appledaine", the corner where the course dodges around a cottage, is safest done in second; you might be tempted to use third but a shade of misjudgement here would land you in difficulties for the road is none too wide.

It is all medium going to Greeba Bridge (2nd cog) where an awkward camber merits due care and attention so that you don't lose wheel grip. Then it's full bore, watching out for the last right-hander (quite a handful) which brings Ballacraine into sight. Bottom gear past the hotel, then a quick snick into second which is used to peaking point (somewhere in the region of 85 m.p.h.) round the rising left sweep. Today you go straight through Ballig Bridge in top...in fine weather. If it's wet special care must be given from this point onwards to the top of Craig Willies hill, for water drains from the banks and makes the road surface treacherous. Watch the sweeping left hander after Ballig and, if in top, ease off for it. And drop into second before Laurel Bank which is plenty fast for that gear but too slow if taken in bottom. It is one of those few corners that doesn't suit normal gear ratios and so requires some judgement to take it at just the right speed.

The winding mile or so, past the quarter-way sign to Glen Helen is taken in second and third, with the rider concentrating on jockeying for position into each corner and getting the most out of the engine. You "float along", as it were, not peaking in any gear but making full use of what width there is in the road. This is not a place to try passing.

The Glen Helen left-hand corner I take in bottom. There is always a spring trickling water on to the exit line and I like to play safe, going slower than it is possible to take the corner but making sure I can dodge the wet patch and come out well positioned for violent acceleration to the right sweep higher up.

Having breasted Craig Willies you get a brief rest along the rising stretch ot the Cronk-y Voddee Chapel cross-roads and the right and left downward swerves which follow can, by concentration and using every inch of road, be taken flat in top --- but only just. The procedure is not recommended for beginners. If you keep it going through this section you feel pleased with yourself but a "chop" makes it seem dead easy and you kick yourself for losing valuable time.

Handley's Corner is an orthodox left-right necessitating second gear. The medium left hander just past the Barregarroo cross roads is taken at "full chat" in top (although there isn't much room to spare) if you want to take full advantage of what is, on a good day, about the fastest part of the course, the drop down to the bridge by the 13th milestone. People think that the gentle left at the bottom of the hill is deadly, but actually it is good for full bore ---125 m.p.h. – if taken on the right line and it ought, I imagine be one of the best spots from which to spectate.

It is very tempting to have a "bash" at Kirkmichael corner in third but wiser to use second and wind it on as you leave the bend so as to waste no time in getting to the next hazard which is Birkin's Corner (15th milestone). Here I don't know what to say. I *have* done it "flat" – but I don't like to! Also it seems to have become more difficult since the war, perhaps because it is bumpier. Anyway it's a borderline case where a "shake of the bottle" probably loses you little more than the satisfaction that a clean sweep can bring.

It is now "set fair" for a full mile to Ballaugh, except for one right hander which is liable to loom up a bit sharper than the rest. Ballaugh Bridge needs bottom gear and there is very little else you can do about it. It doesn't pay to leap long distances and the main object is to have the machine upright as you accelerate violently through the village.

The Quarry Bends? When you have navigated the fairly sharp right hander near the 18th milestone, drop into third and accelerate through the left, right, left swerves that follow. Artie Bell tried the whole lot in top last year --- on a "350"--- but he reckoned afterwards that that caper was "out"

The Sulby "Straight" starts immediately after the last Quarry bend and you get nearly a mile of respite with speeds up to 120 m.p.h. before it is "flaps down" and fine pitch for the right-angle over the bridge. The inviting sweep left after the Ginger Hall Hotel is one of the most misleading sections of the course – as skid marks in queer places at the end of a day's racing will always show. It can't quite be done in second gear stride. If you don't ease off you will find yourself in an embarrassing position on the pavement. And in another mile there is an abrupt left --- I believe it is called Kerroo Mooar --- that most certainly needs caution and second gear, for the place possesses a tree, endowed apparently with the ability to move itself right into your path.

Except for the sharpish left at Glentramman it's all third and top work with swinging lefts and rights into Ramsey. A nasty hump in the road before the school means a momentary flick into third. If it was smooth here you could take the bend unchecked; as it is you have just time to get back into top before anchoring up for Parliament Square --- in which process allow for the wind. It blows in varying strength up the street and unless you are prepared for its vagaries you can easily arrive in the Square going ten miles an hour faster than you meant to.

The Square is crossed in first gear, engaging second by the cinema and using third to May Hill Corner which, now that it has been widened and resurfaced will allow for second. Stay in this gear until after the left-hander which preludes the third gear ascent to the Hairpin. Leave the power turned on until pretty late for being steeply graded it is an easy pull up.

Plenty of clutch slip is called for in accelerating up the rise after the Hairpin but keep an eye open for the medium right-hander that comes just before Waterworks Corner --- it is very bumpy and could easily put you off course for Waterworks itself, which is taken in bottom. Thence it is a matter of ringing the changes on second and third to the Gooseneck. --- hard driving but not quite peaking. Make the engine do the donkey work.

The most common mistake made at the Gooseneck is accelerating fiercely in bottom gear before the machine is upright. I make a very sharp turn of it leaving the acceleration until I am pointing in a straight line ahead up the hill. With a plain pipe one could drive around but megaphones on the outside interfere with the camber.

It is "all out" now up the Mountain, mostly in third gear to the 27th milestone but with a slow down through the "S" bend, formerly called "The Cutting" and now marked by the Guthrie memorial. For me this is a doubtful section and a lot depends on the way of the wind whether I use first or second --- if conditions are adverse second may prove so slow as to bring the engine off the "trumpet".

Over the Mountain Mile one can usually use top gear. In 1938 I could reach about 115 m.p.h., but again it depends on the weather. Watch the r.p.m. and if there is the slightest suspicion of a headwind I drop into third. The Mountain Box is an orthodox left-hander requiring second, while the Stonebreaker's Hut allows third or top, again depending on the strength of the wind and visibility. The series of right-hand sweeps called the "Verandah" is actually treated as one long continuous curve, usually taken in third with a possible brief snick into top, but the left-hander over the stone bridge will bring you back to third – and a cautious one at that. Stay in this gear for the remainder of this stretch but be ready to drop to second for the Bungalow bend.

Approach this section on the slow side, for the whole of the proper negotiation of the "S" bend round the hotel is wrapped up in getting the correct line into the initial left-hander. You can get third for a short spell to the Mountain Gate's left sweep (second) and then the three lefts before Windy Corner allow top on a clear day. They are treated in the same way as the Verandah and if you have everything weighed up just right you can get through flat out. If not you will have to "blip" it to keep on your line.

'Windy" needs second gear and some caution. You can see round it alright, but it is downhill here and you may find yourself approaching too quickly. Nothing counts then until the double left at the 33rd milestone. This was a "very nearly flat out" proposition before the war and the recent alterations have not made it any easier. The approach and exit are faster but the surface is loose and treacherous so it is still of the "just off flat" variety.

The subsequent section is taken at a fair clip in third until the approach to the left-handed Keppel Gate looms up and demands an immediate drop to second, followed by an equally smart re-engagement of third for the descent to Kate's Cottage, which is best tackled still accelerating in this gear. It is a handful --- much bumpier than it was --- and there is a strong temptation to tweak the "throttle". Changing up too early is definitely not recommended for the engine is on peaking point and the act of swapping gear may spoil your line. So stay in third until you are safely round and then tuck in for the thrilling 120 m.p.h. dive on to Craig-ny- Baa.

It is surprising how soon it is over and before you have time to recover your breath it is "let go both anchors" and rapid back-pedalling on the gear lever in readiness for the sharpest pull-up on the whole course. It is not wise to leave these operations until after you have passed the road traffic warning notice board, but do not rely on this mark. On race day, with many spectators crowding the banks you may not see it at all. In any case I don't believe in riding by lamp-post and suchlike land marks.

Keep well to the left of the Craig on the approach and make a sharp turn of it so as to be well placed for a straight line getaway down the hill to Brandish Corner which is a fast second gear bend. Hillberry, before it was redesigned was a second gear job;. now it's third which can be held through the sweeping lefts over Cronk-ny-Mona, treated Verandah-fashion as one long bend.

Look out for "Signpost" and get into low gear early. It's downhill and a trip up the slip road wastes so much time!

Bedstead Corner is a nasty, bumpy affair where you can do one of two things --- either a lot of revs in second or a modest number in third. I really don't know which is best. The sharp right-handed "Nook" will not allow anything higher than first and the short spell in second which follows must be kept short. If you aren't back in bottom gear and beginning to brake before you sight Governor's Bridge you will be too late to make a close-in circuit of the hairpin. Use plenty of clutch slip and revs in making the turn, but remember the loose gravel in the dip and also the steep camber on the exit. Rather than attempt any great hurry through this section it pays to concentrate on getting a good line out of the hollow for a smashing exit in second --- paying due regard to the weather for the surface of the Glencrutchery road can be very slippery when wet.

So, in 24 mins.52 secs. we are back where we began. Some day, someone will clip 133 secs. off that figure to put the lap record where, but for the war it might well be today ---at 100 miles per hour.

Part Two

Profiles of their post-war works riders
from 1947 to 1955

Serafini's emphatic Gilera victory at the 1939 Ulster Grand Prix brought down the curtain on an absorbing decade of racing. By then BMW, Gilera and DKW had demonstrated the absolute superiority of multi-cylinders and supercharging. Norton chose not to supercharge their single-cylinder design although Vincent HRD and other British manufacturers did investigate the possibilities. In the 1937 TT series Phillip Vincent and his chief designer Phil Irving made a bold foray into forced induction with their TT Replica single. The exercise brought nothing but frustration which saw them revert back to the normally aspirated model.

Despite Vincent's lack of success, two other firms were prepared to challenge foreign superiority. AJS had demonstrated just how fast their complex V4 could be if only they could improve on its reliability. Velocette's blown twin was a more refined design housed in a frame that provided superb handling unlike the Ajay whose handling was atrocious. The 'Roarer' as the Velo was known made a brief appearance in a practice session at the 1939 TT.

Stanley Woods described it as the finest handling bike he had ever ridden and having the speed potential of its rivals, once its teething problems were sorted. The 'Roarer' was judged to require further development and due to the war it was never raced in anger.

Following the end of hostilities road racing was resumed in a tentative manner. With labour and materials in short supply and petrol strictly rationed it was a remarkable feat for the Manx Motorcycle Club to re-stage the Manx Grand Prix in September 1946. The same could be said for the Auto Cycle Union in staging the 1947 TT series, particularly in the extreme austerity of the times.

A severe handicap was the low octane fuel available at the time but more significantly supercharging was banned by the F.I.M. the governing body of motorcycle sport. In effect Norton was given a reprieve, an opportunity to regain their pre-war superiority in the short term at least.

Their immediate priority was to modify their motors to perform on the dreaded 70 octane 'Pool' petrol. Lower compression ratios meant slower lap times which were forcibly demonstrated at the 1947 TT series. The evergreen Harold Daniell led a Norton 1-2 in that year's Senior at an average speed of 82.81 mph. In winning the 1938 Senior Daniell averaged 89.11 on what was essentially the same machine. It demonstrated the radical effect of inferior fuel on racing motors.

Daniell was Norton's sole representative from their pre-war team. Fred Frith was entered for the 1947 Senior on a Moto Guzzi which was basically a replica of Stanley Woods 1935 Senior winner. Frith became a non-starter following a practice crash at Ballacraine. He returned to racing at the end of the season and on Velocettes, until his retirement in 1949 he was a consistent winner and also the inaugural 350c World Champion. John White had married and with a young family he decided that racing was not appropriate for a married man.

From 1947 until their official works team was disbanded in 1955 Norton supported racing to its fullest extent. Behind the scenes was the 'Professor', the taciturn Joe Craig extracting the last ounce of horsepower from his beloved Nortons and ruling his team with quiet authority. Their list of riders reads like a 'who's who' of British and Commonwealth racing talent. Other British riders did reach the peak of their profession without an involvement with Norton but these were few indeed. The popular Les Graham who became the inaugural 500cc World Champion as an AJS work's rider is one example. Bill Lomas was another. He rode a variety of makes during his career, the highlight of which was gaining the 350 World Championship on a Moto Guzzi in 1955 and again in 1956.

Some Norton riders, like Geoff Duke and John Surtees were regarded as the greatest of their era, the superstars of the 1950s and 60s. Others; like Ireland's Artie Bell and Rhodesia's Ray Amm left an indelible impression on the racing scene before their star was extinguished. For some, their association with Norton was on a casual basis, in the manner of Irishmen, Louis Carter and Manliff Barrington who represented them in local races.

Following his 1951 Senior Manx win, Dave Bennett was on the brink of a promising career as a Norton teamster only to crash fatally at the 1952 Swiss Grand Prix. Veteran pre-war rider Sid Lawton became a member of the 1953 team. He won the 500 class of the North West 200 but at the TT he suffered horrendous injuries in a practice crash and never raced again. Australian rider Gordon Laing was less fortunate; when after a good performance at the 1954 TT series he was given a works ride in the Belgian Grand Prix only to crash fatally in the 350 race. Those golden years of road racing provided moments of unforgettable excitement and drama; with a blend of tragedy that comes with such a demanding pursuit. The Isle of Man TT was the main focus of the racing calendar in that era. It no longer features in today's World Championship; being classed as far too dangerous for Grand Prix events. Times have changed but for riders of that Golden Era nothing ever compared with a win at the Island.

The Manx races were unique in that there was no massed start. Riders were dispatched at intervals; sometimes in pairs and at other times individually. The situation became a race against the clock; as your closest rivals may have started ten minutes or so ahead of or behind you. Consequently it was unlikely that you ever saw them until the race was over. Success in the Isle of Man demanded great concentration and discipline from the rider. Another factor was an efficient signalling system, not just from your pit crew but also from your personal timekeepers stationed around the course.

A circuit more demanding or difficult than the Island would be hard to find. The old Nurburgring, situated in the Eiffel Mountains was equally daunting and both venues were subject to inclement weather conditions. A single lap at the Isle of Man mountain course covers 37 miles or 58 kms of public roads, which for the newcomer was a bewildering experience. A generally accepted fact was that it took several visits to memorize the multitude of corners and landmarks. This also involved the negotiation of several villages and the ascent and descent of Mount Snaefell, which is often shrouded in mist. The road is generally quite narrow and does not afford the luxury of run-off areas in the event of a mishap. Over the years its surface was gradually improved and with those improvements came a marked increase in lap records.

The remarkable Jimmie Simpson became the first rider to lap at 60, 70 and 80 miles per hour. When one considers the less than perfect state of the roads in the twenties and thirties, these were remarkable achievements. In 1938 Harold Daniell recorded a lap of 91 mph in the process of winning the Senior TT of that year. Due to the war his record stood until 1950 when rising star Geoff Duke eclipsed Daniell's pre-war time with a lap at 93.3. In both instances they were on work's Nortons whose power output would have been markedly similar, despite the passage of time. It was the revolutionary 'Featherbed' frame, plus Duke's brilliance that made the crucial difference. In the following year Duke scored a Junior/Senior double; with a record lap of 95.22 in the Senior.

The possibility of registering a magic 'ton' at the Island was beginning to appear feasible, although unlikely that a single cylinder machine would be the first to achieve that goal although the wild man from Rhodesia, Ray Amm recorded a blistering 97.4 mph lap in winning the 1953 Senior TT. Duke had switched to Gilera in 1953 however it was not until 1955 that he was in a position to be the first to top the magic ton. He led that year's Senior TT from the start and on his fourth lap he came tantalizingly close to that 100 mph lap. There was great excitement when the timekeepers awarded him the honour but this was later revised and his time was officially recorded as 99.97 mph. The elusive 'ton' was finally realized at the 1957 Golden Jubilee TT. That great Scot, the late Bob McIntyre became the first rider to officially break the 100 mph barrier. Geoff Duke recommended McIntyre as his Gilera replacement, following Duke's accident at Imola earlier in the season. McIntyre rose to the occasion and scored his own Junior/Senior double and in the process he put in a record lap of 101.1 mph. Sadly Bob was killed at a wet Oulton Park meet in 1961. Ton-up laps became the norm in the years that followed although it was the quiet Scot who became the first.

Ray Amm (1952-1954)

Ray Amm

1953 Senior TT

The British Commonwealth made a great contribution with riding talent and of all the 'colonial boys' none were wilder than William Raymond Amm. He began his racing career in 1949 on the grass tracks of his native Rhodesia. In 1950 he progressed to road racing and in that year's Port Elizabeth 200 Handicap he finished third on a Norton. He also campaigned a Grand Prix Triumph on the local circuits and these experiences persuaded Amm to travel to Europe, determined to prove that he could match it with the world's best.

Accompanying Ray was his wife Jill, his loyal assistant and most supportive fan. All their resources were invested in a brace of new Manx Nortons and the ubiquitous old van. There was no turning back; Amm had to succeed in Europe or starve. His initial racing excursions demonstrated that he was more of a grass-tracker than road racer. But there was no doubting his determination. His first Isle of Man venture was the 1951 TT. In the Junior TT Amm finished a creditable ninth at 85 mph and despite problems in the Senior he was 20th at 78.42 mph.

Amm's lurid style did not compare with the elegance of Geoff Duke or his contemporaries although his courage was never in doubt. It was a factor that saw him drafted into the Norton team for 1952, partnering Duke and Reg. Armstrong. Amm's Junior TT ended with a crash on lap 3 and in the Senior he finished a strong third behind Reg. Armstrong and Les Graham on the MV Agusta.

Amm performed some heroic rides in the continental races, backing up team leader Duke in their attempts to stay with the faster Gilera fours. Following Duke and Armstrong's move to Gilera at the end of the season, Amm was promoted to Norton No.1, supported by veteran Jack Brett and Australian Ken Kavanagh. Amm's elevation to team leader was amply vindicated at the 1953 TT series. Supported ably by Kavanagh, Amm headed a Norton 1-2 in the Junior event, with Amm also recording fastest lap. In 3rd place was Fergus Anderson on a 320cc Moto Guzzi. Anderson's bike was basically a stretched 250, which was developed into a world-beating 350 that gained every World Championship from 1953 to 1957; a portent of Norton's eventual decline in that category.

Friday's Senior proved to be one of the most dramatic ever. Clear race favourite was Geoff Duke, making his Gilera debut in the Island. His strongest challenge was expected to come from ex-AJS World Champion Les Graham who in 1951 switched to MV Agusta. Norton could not be discounted either, despite the speed advantage of the Italian fours.

Duke led from the start, raising his 1951 record lap of 95.22 mph to 96.38 mph from a standing start. In second place was Les Graham; then no sooner had he begun his second lap that he crashed fatally at the foot of Bray Hill. Nothing could restore the vanished joy of a day, which began so brightly. Meanwhile the race continued. Amm had started the race at 61 and Duke was 67, so they were virtually together on the road at that stage. Duke was briefly hampered by gear selector problems, allowing Amm to overtake him. He later recalled that it gave him an opportunity to study Ray Amm at close quarters; an experience that he described in his autobiography as 'frightening'.

Events took a dramatic turn on the fourth lap when Duke slid to earth at Quarter Bridge and damaged the Gilera too severely to continue. Despite a minor tumble on the final lap, Amm raced on to an emphatic win, with a record lap at 97.41 mph. In second place was Jack Brett with Armstrong third on the second Gilera. The flying Rhodesian had joined that elite band of double TT winners. He did not feature in other classic Grands Prix during 1953 after crashing heavily in the 350 French Grand Prix and breaking a collarbone.

Amm remained loyal to Norton for the 1954 season, supported by Jack Brett and 1953 Senior Clubman's winner Bob Keeler. Amm was poised for a win in the Junior TT when surprisingly he retired on the fifth lap with mechanical problems.

His retirement enabled New Zealander Rod Coleman to score a popular win on the three-valve works 7R. AJS fans were all agreed it was long overdue, and their first since Jimmie Guthrie's win in the 1930 Lightweight.

Friday's Senior proved to be the most controversial race of the 1954 season and possibly the most controversial TT ever. Bad weather delayed the start, and unlike the 1935 Senior this one was not postponed until the Saturday. On the opening lap Duke led Amm by 14 seconds at 88 mph; a commendable speed, considering the adverse conditions and as the race continued they deteriorated even further. Duke stopped for fuel at the end of lap 3, allowing Amm to gain a considerable lead. Amm's strategy was to go the seven-laps distance non-stop. By lap 4, general opinion was that the weather was clearing but the race stewards decreed otherwise and made the decision to stop the race once the leader completed four laps. As a result Ray Amm was declared the winner, with Duke second and Brett third.

In addition to his dramatic TT victory Amm scored a Junior/Senior double at the Ulster Grand Prix. His final Norton appearance was at Aintree and in his usual lurid style he crashed spectacularly during the 500 race. Norton's policy for 1955 was to drop 'works' machines in favour of development versions of the standard Manx racers. Amm was faced with another season on outmoded machinery or a move to MV Agusta. He chose the latter, with a debut at the Imola circuit in April.

Amm was not enamoured with the MV, particularly the 350 which was no match for the ultra-light Moto Guzzi singles. He used his own Norton for some of the official practice periods and inexplicably he fitted the worn tyres from the Norton to his race MV. The 350 race started at a furious rate with Amm in third place behind the Guzzi pair of Kavanagh and Sandford. Perhaps Amm was pushing too hard for he crashed early in the race. At first it did not appear to be too serious but on this occasion Amm's run of good fortune came to an abrupt end.

He was flung down a grass bank and appeared to be in fair shape, then his head struck a submerged steel post with fatal results. Imola's surface was quite abrasive and on inspection the MV's tyres were found to be completely bald. Ray Amm's departure from racing was as spectacular as his arrival. Tragic as it was Amm's death came as no real surprise, riding as he did on the brink and at times beyond it. A fellow rider once commented on the paper-thin leathers he wore. His response demonstrated his fatalistic approach to racing with the reply that 'when he did go it won't matter what leathers I'm wearing'.

Reg. Armstrong (1952)

Reg. Armstrong

1952 Senior TT

In a career that began modestly enough in 1946 in Irish road races, Reg Armstrong campaigned a wide variety of makes. He was born in Liverpool in 1926 where his parents had settled temporarily. The family returned to Dublin in the early thirties and Armstrong Senior became a successful motor dealer. Reg's racing debut was at the Bangor Castle races where he rode a pre-war Norton International in this and in other Irish events during 1946.

To further Reg's career his father purchased a brand-new 350 Manx Norton for him to contest the 1947 season. Success in local races proved elusive but with a 500cc motor borrowed from Artie Bell and fitted to the 350 he was entered for the 1947 Senior Manx Grand Prix. His other entry was a 250cc Excelsior Manxman in the Lightweight category. He was well placed in that event until his exhaust came adrift, putting him back to fifth at the finish.

Reg's satisfaction at finishing his first Island race was marred by the death of his friend Benji Russell. He was a protégé of Stanley Woods who provided a 250 Moto Guzzi for Russell's Lightweight entry. The promising newcomer was leading the race when on the approach to Ramsay on his third lap he hit a manhole cover, crashed and was killed instantly. Deeply grieved with Russell's death, Armstrong withdrew from the Senior.

On the Excelsior he was leading the 250 class of the 1948 North West 200 when he retired with mechanical problems. For the Skerries 100 he was loaned one of the newly introduced AJS 7Rs and having a reliable and good-handling bike was a great confidence booster. He was entered for the 1948 Manx Grand Prix with a 7R in the Junior and a Grand Prix Triumph in the Senior. Armstrong's improved performances caught the attention of the AJS competition department and following the Manx he was offered a ride on an AJS 'Porcupine' twin at the Ansty aerodrome circuit. Compared to his previous mounts the 'Porc' was a potent device and Reg did well to finish fourth in his first race for AJS.

His reward was a place in the AJS works team for 1949. Other members were Les Graham, Ted Frend and new boy Bill Doran who finished in second place in the 1948 Senior TT on his Norton. Reg's contract was to ride only the 7R; perhaps there were not enough 'Porcupines' to go around. He proved himself as a reliable team man with good placings at the TT and also the classics, amassing enough points to finish second to Fred Frith in the 350cc World Championship.

For the 1950 season Armstrong was entered on Velocettes provided by entrepreneur Nigel Spring. In 1949 Spring was Frith's entrant in his successful bid for the World Championship. Armstrong had several good results on the continent but was unable to match the veteran Bob Foster who became 350 World Champion for 1950, also on a sponsored Velocette. This was also the debut of the MV Agusta four. On two occasions; the Dutch TT and Belgian Grand Prix, Reg had memorable outings on the new model whose handling proved challenging and its general performance mediocre. The new model took two years of sustained development mainly by Les Graham to make it competitive. Just when his efforts should have been rewarded he was tragically killed at the 1953 Senior TT.

Reg re-signed with AJS for the 1951 season to ride both the 7R and the 'Porcupine'. It proved to be a generally unfruitful year for AJS. Their best result was Doran's second place in the Senior TT, plus his 350 win at the Dutch TT.

This had been an unproductive year for Reg Armstrong and then just before Christmas he was reviewing his options when he received a phone call that altered his future completely. It was none other than Joe Craig offering Reg a place in the Norton team for 1952; other members were team leader Geoff Duke and newcomer Ray Amm.

His initial works outings were a mix of success and disappointment. Reg won the 500 class of the Leinster 200, and at the North West 200 he was disputing the lead in the 500 event when his primary chain broke; a bad omen and a portent of his fortunes in the forthcoming TT series. Duke won the Junior TT in convincing fashion with Armstrong in second place and New Zealander Rod Coleman third; his AJS denying Norton a 1-2-3 finish. Friday's Senior promised to be a much sterner contest for Norton and AJS teams; in particular the challenge from MV Agusta. Les Graham's tenure with MV had been dogged by problems that would have dismayed most riders. Thanks to his perseverance plus the co-operation of MV engineers the Italian four had achieved a level of reliability to match its undoubted speed.

Nevertheless, Duke was clear favourite for the Senior with his greatest challenge expected from Graham, whose riding skills were rated on a par with Duke, the 'maestro'. The opening lap saw Duke take an early lead over Graham, with Amm in third place and Armstrong fourth. Duke's Norton however was experiencing misfiring due to a weak mixture but despite the handicap he still maintained his position.

Sensation on lap 5! Duke retired at the pits with an inoperative clutch; giving the lead to Graham. But he too was experiencing problems: the MV developed an oil leak, saturating his right boot and gear change; resulting in a missed gear change: the revs soared, damaging the valves and taking the edge off the MV's performance. Armstrong responded to the challenge and took the lead on the final lap, increasing it to 27 seconds at the finish, but the drama was unfinished, when at the precise moment he crossed the finish line his primary chain broke; it was Friday the 13th too! Reg Armstrong had joined that elite company of Senior TT winners, amply justifying his place in the Norton team.

Following the TT, Reg scored a first in the 500 class of the German Grand Prix and secured places in the Dutch and Swiss events. At that stage he was leading the World Championship by three points over Gilera star Umberto Masetti and Les Graham but there was no Luck of the Irish at the Ulster Grand Prix where he retired with a broken primary chain.

Masetti and Graham both had retirements at the Ulster, leaving Armstrong still holding his three-point lead. A good placing at the Italian Grand Prix would see him as World Champion. It was not to be. His sixth place was not sufficient to defeat Masetti who became 500cc Champion for a second time. Despite heroic rides by Duke and Armstrong it was obvious there were scant rewards in racing an obsolescent single.

In 1953 Reg moved to Gilera as back-up rider to Duke, leaving open his options to ride in other categories. NSU were developing a competitive 250 to challenge Moto Guzzi's monopoly of that class. Riding an NSU, Reg won the Swiss Grand Prix and was runner-up to teammate Werner Haas in the 1953 World Championship. His three years with Gilera showed his worth as a reliable team man, finishing second to Duke in the 1953 and 1955 World Championship.

Prior to his retirement in 1956 he won his last race, a non-championship 500cc event in Sweden. He maintained his association with NSU and became Irish Importer for NSU motorcycles which at that time were Europe's largest motorcycle manufacturers. It was then they took a quantum leap and put all their resources into the R80; the first production rotary-engine car. It proved to be under-developed and was so beset with warranty claims that it bankrupted many dealerships and eventually its manufacturers suffered the same fate.

Armstrong also secured the Honda motorcycle agency for Ireland; a fortuitous move, and one that elevated him into the millionaire bracket. After defying the odds in the top echelon of road racing for so many years Reg Armstrong died in a car crash near his home in 1979.

Artie Bell (1947-1950)

Artie Bell

1948 Senior TT

Artie Bell's racing career is a tragic case of 'what might have been'. Only for team orders he may well have won the 1947 Senior TT in his first-ever Island appearance. Had there been a 500cc World Championship in 1947, Artie Bell would, with mathematical certainty have won it. Again in 1948 Bell would have emerged as 500cc Champion.

In a post war career that spanned four seasons Artie Bell rode Nortons exclusively, although it must be said that the 'Garden-gate' Norton of that era could be an absolute brute; a device that required sheer strength and determination for success. The introduction in 1950 of the 'Featherbed' frame at last offered Artie Bell a machine worthy of his ability. Then at the Belgian Senior Grand Prix he was involved in a horrendous pile-up and Bell's career was over.

Arthur James Bell was born in Belfast in 1915 and from an early age he developed an interest in all things mechanical. In 1937 despite parental opposition he rode in local handicap road races. He was moderately successful in his racing forays and also competed strongly in trials. Racing came to a halt in Ulster at the outbreak of World War 2. During that period Bell was involved in aircraft production at Belfast's Short and Harland works and while there he met the innovative engineer, Rex McCandless. Immediately after the war they went into partnership in a heavy machinery business which developed into a highly successful operation. Bell first came under Joe Craig's scrutiny at the 1947 North West 200. His win in the Senior event was enough to convince Craig to sign him up for the 1947 works team; almost a case of Bell being thrown in at the deep end

Apart from one visit to the Isle of Man in 1935 as a spectator he had absolutely no experience of riding there. Accompanied by his wife Iris, Bell went to the Island two weeks before official practicing commenced. In a fortnight of intensive note making and research and riding the course, he applied himself to learning the most demanding circuit in the world.

Regardless of his lack of experience Bell led the 1947 Senior for three laps. When he pitted for fuel Craig ordered him to ease his pace. Artie duly obeyed; allowing Harold Daniell to come through and win his second Senior. Bell's second place and the share of fastest lap in his very first Senior were most commendable.

His Continental forays were equally rewarding, with first place in the Dutch TT and Ulster Grand Prix, plus second in the Belgian and Swiss Grands Prix. Perhaps it was ironic that no World Championship existed in 1947, for A.J. Bell would have been a clear winner. In 1948 Bell had the satisfaction of a win in the Senior TT, a race that was memorable for its high attrition rate. Apart from the Norton and AJS works teams there was a strong challenge from Moto Guzzi, plus an entry of 6 of the newly introduced Grand Prix Triumphs. Added to these were the private entrants, making a total of 55 starters which was slightly less than average for a TT race.

Many of the stars fell by the way in their attempts to match the pace set by Tenni on his flying Guzzi in the early stages. After heading the field for four laps and making fastest lap at 88.06 mph he was beset with problems and reduced to ninth place at the finish. Taking advantage of Tenni's misfortunes Bell assumed the lead on lap five and held it to the finish. He was the sole British works rider to survive the event which he won by the uncommon margin of ten minutes, ahead of the Nortons of Bill Doran and Jock Weddell. Sad to relate that following the TT the mercurial

Omobono Tenni was killed during practice for the Swiss Circuit de Berne event. With his continental successes Bell would have again been 500cc World Champion with a lead over teammate Johnny Lockett.

Norton did not officially contest the 1949 season; instead they loaned last year's bikes to a syndicate of Daniell, Bell and Lockett with Steve Lancefield supervising their maintenance. Apart from a win in the Senior North West 200 and second place at the Ulster Grand Prix behind Les Graham's AJS, Bell did not feature strongly in other classic events. He experienced problems in both the Junior and Senior TTs, finishing third in the Junior and fourth in the Senior.

In 1950 there were two events that elevated Norton's fortunes. One was the McCandless 'Featherbed' frame which prolonged the Manx Norton's life more than any other factor. Second was the signing of Geoff Duke. Few riders have risen to the peak of their profession so rapidly and so convincingly. His first outing on the new model was at Blandford where he won the 500 event. More success came at the North West 200; with a 350 win for Duke and victory for Bell in the 500.

Norton continued their success at the TT with a clean sweep in both Junior and Senior TTs. In the Junior it was a Bell, Duke, Daniell 1-2-3 and a new lap record to Bell at 86.49 mph. Duke reversed the tables in the Senior with Lockett in third place. In the process of winning, Duke raised the lap record to 93.33mph to finally eclipse Daniell's 1938 record of 91mph..

Next was the Belgian Grand Prix, held on the ultra-fast Spa Francorchamps circuit. In the 350 event the Nortons were outpaced by Bob Foster's ageing but rapid Velocette, with Bell second and Duke third, which was to be Bell's last race finish. The Senior developed into a furious 3-way battle between Bandirola (MV) Graham (AJS) and Bell. At the corner before La Source hairpin all three were doing over the ton when Bandirola suddenly backed off. Taken by surprise Graham was forced to brake violently and was brought down when his front wheel locked. Bell had no option than to take to the dirt verge to avoid Graham. The AJS continued to slide along the road and was hit by Bell's

Norton. Bell disappeared with both bikes in a cloud of dust beneath an elevated observation post. By a miracle Graham was practically unhurt but poor Bell received terrible multiple injuries. For weeks he was in grave danger, and thanks to dedicated medical care and his splendid physique and high courage he made a reasonable recovery.

Artie Bell never raced again and will always be remembered as a man who was never content to accept second place; whatever the odds against him. His TT record was a model of efficiency; where from just 8 starts he scored 2 wins; the 1948 Senior and 1950 Junior, 2 seconds the 1947 and 1950 Senior and 2 thirds; the 1948 and 1949 Junior. His lowest finish was 4th place in the 1949 Senior with just one retirement; the 1947 Junior.

A fascinating exercise is to speculate on how the record books may have read, had Artie Bell's career not come to a premature end. There is no doubt he would have been a serious rival to the aspirations of Norton's new superstar, Geoff Duke. Bell never fully recovered from those injuries, having lost the use of his right arm plus other problems. He still maintained an involvement in racing, and acted as entrant for promising riders He died of a heart attack in 1972, brought on by massive doses of pain killers he was taking. Artie Bell was 57.

David Bennett (1952)

David Bennett

1951 Senior Manx GP

Dave Bennett's tenure as a Norton team member was tragically brief, nevertheless he merits an inclusion in this volume. He was born in Birmingham in 1928, the youngest of three brothers. With Stan and Clive employed at Ariel Motorcycles it was inevitable that Dave would pursue a similar career. His duties at Ariel were mundane to start with and later he became involved in road testing, which proved to be a much more enjoyable experience for the enthusiastic teenager.

On his 18th birthday Dave began his stint of compulsory National Service. After basic training he was posted to Tel el Kebir in Egypt. Off duty periods were occupied with speedway racing on heavily modified army bikes. With his National Service completed Dave returned to his old employer Ariel who supported him in national trials and scrambles. One highlight was the 1949 International Six Day's Trial, held that year in Wales. He was poised to win a Gold Medal but was forced to retire on the fifth day.

Despite his trials success, Bennett was determined to move on to road racing. In late 1949 he secured a position at Norton where his growing reputation earned him a place in the race shop. It was satisfying work but the low wages ruled out any hope of purchasing competitive race machines. It was the intervention of family friend Ernie Earles that hoisted Dave on the first rung of the racing ladder. Earles provided a 350 and 500 Manx Norton for

Dave to ride in the 1950 season. The new 'Featherbed' model did not become available to private owners until 1951, so for the 1950 season it would have to be the outmoded 'Garden Gate' version for Bennett and other aspiring riders..

He campaigned the Nortons at a variety of circuits during the year; the high-light of which was the 1950 Manx Grand Prix. The newcomer featured regularly on the practice leader board and on race days he was a model of consistency. Veteran Don Crossley on a 7R AJS won the Junior after a spirited tussle with Peter Romaine and Robin Sherry and almost unnoticed after a steady ride was Dave Bennett in 8th place. Favourite for the Senior was Peter Romaine on the Beart-prepared Norton. He held off various challenges to lead throughout. In second place was Mick Featherstone with Harold Clark third and after another steady ride was Dave Bennett. Norton machines occupied the first fifteen places.

Following his encouraging Manx performances, Bennett's main focus for 1951 was the Manx Grand Prix. Once again Ernie Earles provided new bikes, in this instance a brace of 'Featherbed' Nortons. Despite Bennett's connection with Norton's race shop, both were completely standard and not thinly disguised specials. During practice for the Manx, Bennett featured strongly in both 350 and 500 sessions. Favourite for the Junior event was Robin Sherry on a rapid 7R AJS which in all probability was a works special.

Bennett made a strong challenge to Sherry throughout the race, holding second place until the fourth lap when he retired with a broken primary chain. Inclement weather forced an overnight postponement of the Senior. Race day was sunny but strong winds could prove bothersome at exposed parts of the circuit. From the start Bennett went into an early lead which he maintained throughout. In second place was Don Crossley on a Beart-Norton and in third was Robin Sherry on a prototype Matchless G45 twin. Sherry was displaced at the finish by veteran Denis Parkinson on one of the swarms of Nortons that followed the leader home

Following his record-breaking Manx win Dave Bennett had proved himself worthy of a place in the 1952 Norton works team. The newcomer would be partnering Geoff Duke and Reg Armstrong, although his first works outing was the 1951 autumn race meeting at the Thruxton circuit. The 'Daily Express' Trophy final saw a titanic battle between Les Graham's MV and Dave Bennett's Norton, with the MV edging out Bennett's Norton and John Surtees' Vincent Grey Flash.

The opening Grand Prix of 1952 was the Swiss; with the Norton team on the line against a strong AJS works trio, plus the Gilera and MV entries. During practice it became apparent that the local fuel was not compatible with the works Nortons. Bennett's motor was affected more than his teammates' and required a complete re-build. It was to have disastrous consequences for Bennett.

The 500cc GP began at a furious rate and saw Duke holding first place ahead of Graham, Brett and Coleman with Bennett in mid-field. It soon developed into a race of attrition. Duke's Norton, which was not the subject of a re-build as was Bennett's, succumbed to the damage caused by the faulty fuel. As the race progressed there were further retirements and on lap 22 of the 28-lap race the order was Doran, Bennett and Brett. His was the lone works Norton battling with the more experienced AJS duo of Doran and Brett.

This developed into a fierce battle for the lead which changed repeatedly and with no quarter given. As Doran and Brett sped past the pits on lap 27 there was no sign of Bennett. It was at first presumed that he too had suffered engine trouble but in his efforts to stay with the AJS pair he had run out of road, hit a tree and was killed instantly. It was a tragic circumstance. Only for that engine re-build following practice Bennett's Norton may well have expired as did Geoff Duke's, thereby saving him from that fatal crash. It was not to be. A saddened Brett and Doran crossed the line in that order. Thus ended the all-too-brief career of a rider whom Geoff Duke described as having 'exceptional talent'.

Kenneth Bills (1947)

Kenneth Bills

1947 Junior TT

World War 2 had the effect of curtailing the careers of so many riders. At the cessation of hostilities those riders able to, went on to resume their interrupted vocation. Freddie Frith and Harold Daniell were two stars who went on to major achievements in the immediate post-war years.

One rider who was denied a more rewarding career was Kenneth Bills. An optician by profession, Ken Bills' first Island venture was the 1934 Manx Grand Prix. On a 250 Rudge he finished in fifth place in the Lightweight with a seventh berth in the Senior on a Vincent HRD. Bills persevered each year at the Manx and in 1938 he was rewarded with a Junior/Senior double, riding machines prepared by Norton specialist, Steve Lancefield. In more peaceful times a win in the Manx invariably led to a place in the Norton works team.

Unfortunately for Bills, Europe was building up to a world war and besides, Norton were heavily involved in war production and not officially supporting racing in 1939. Tensions in Europe were inexorably building up to crisis point and somewhat optimistically the Manx Motorcycle Club pressed on with plans to run their Grand Prix in September. On 1 September Germany invaded Poland; Britain's ultimatum to Germany expired on 3 September. As a result Britain declared war on Germany. Like many of his contemporaries, Ken Bills joined the RAF and qualified as a pilot.

After a lapse of 7 years and a concerted effort by the Manx Motorcycle Club racing was resumed in 1946. Some entrants were able to purchase what were virtually brand new Manx Nortons. In effect they were part of a batch of 1939 International Grand Prix models that were shipped as spares to the Isle of Man for the 1939 Manx Grand Prix. With the races cancelled the bikes were never used and were providentially stored out of harm's way in the Island.

Practice sessions for the 1946 Manx were generally restrained affairs. With vital components like spark plugs in short supply it was not a time for chasing lap records. The weather was also un-co-operative, with rain and mist for much of the practice sessions.

Ken Bills showed that he had lost none of his pre-war brilliance with a popular win in the Junior. If the weather was inclement for the 350 race then the Senior was run in diabolical conditions. Dark horse for the race was Irishman Ernie Lyons, entered on a prototype Triumph twin. Mindful of the treacherous conditions and blissfully unaware of a broken frame he rode a masterly race to finish in first place ahead of the Nortons of Bills and Manx stalwart Harold Rowell.

Bills' impressive Manx record (3 wins, 3 seconds and a third spot) all on Nortons, secured him a place in the 1947 works team. The bikes were virtually the same as those campaigned by the 1938 team and required modifying to cope with the dreaded 'Pool' petrol. Craig's priority was to concentrate on the more prestigious Senior category. This was borne out at the 1947 Junior where Velocette scored an emphatic 1-2-3-4, headed by veteran Bob Foster. Bills' was Norton's lone work's finisher in a lowly 21st position. His Senior mount expired on the last lap at Union Mills, his engine suffering from the effects of Pool. It was an inauspicious start to a season on works machinery, however his fortunes were about to improve markedly.

Spa Francorchamps was the venue for the 1947 Belgian Grand Prix and it was a foregone conclusion that Britain would dominate the Junior/Senior categories. It was in the 350 race that Norton sprang a surprise. After Norton's poor showing at the Junior TT it was expected that Velocette would continue their dominance at Spa. It proved to be Ken Bills' moment of triumph, leading the race from start to finish. That particular 350 proved to be an absolute flier and Norton teamster Ernie Lyons rode it in the 500 race and finished 8th, making him the first 350 rider home. The 350 Belgian was Bills' sole classic victory on a works Norton and until his retirement in 1950 he switched his allegiance to Velocettes in the 350 class and occasional outings on Triumphs in the 500 category. That was his format for the 1948 TT; a Velo for the Junior and a Grand Prix Triumph for the Senior, disappointingly neither went the distance. The Senior was remembered for its high attrition rate with just 23 finishers from 55 starters. Of the six GP Triumphs that contested the Senior, not one reached the finish line.

There were improved performances on the continent, particularly at the Belgian Grand Prix. Foster headed a Velocette 1-2-3 in the 350 race with Bills in fifth position behind Lockett's Norton. Five GP Triumphs started in the 500 event, won in stylish manner by Norton teamster, Lockett. Unlike the TT, every Triumph finished the race with Foster, Whitworth and Bills in 4th, 5th and 6th places. There was a better result for Bills at the Dutch TT where he scored a fine win for Velocette in the 350 event, after a race-long duel with Freddie Frith.

Bills' 1949 season began at the North West 200 in Northern Ireland. An enormous field faced the starter for the 350 race, which saw Frith and Bills make a smart getaway to break clear of the pack. As the race progressed the Velocette duo began to lap the back markers. It was during an overtaking manoeuvre that Bills tangled with another rider and crashed out. His injuries were serious enough to bar him from the TT and other classics.

His swan song was the 1950 season which coincided with Norton's dominance of the Junior and Senior TT. Following Bob Foster's early retirement in the Junior TT, Velocette were never in the picture. Ken Bills' was the first Velo to finish; albeit in 9th place ahead of Norton stalwart Harry Hinton and Dave Whitworth in 11th on another Velo. Sad to relate, Whitworth was killed at the Belgian Grand

Prix on the following weekend. That great enthusiast was a popular member of the continental 'circus' and for 1950 he elected to confine his continental racing to entries in classic GPs like the Belgian. Like so many other riders his career was interrupted by the war. He returned to racing immediately after the war, and could well have won the 1947 Junior TT. His 2nd place plus fastest lap behind Bob Foster proved to be his best-ever Island finish.

There was a change of circumstances with Ken Bills' original Triumph entry in the Senior. In a bold move the Vincent factory entered their chief tester George Brown on a 500 Grey Flash, a race version of the Comet roadster. Brown had not fully recovered from the effects of an earlier crash at Eppynt in Wales and felt unfit to practice. Bills took over Brown's entry and finished in 12th place, just missing out on a First Class replica. It was still a creditable performance on a machine that was more suited to short circuit racing, rather than an endurance event like the Isle of Man. Perhaps it was fitting that his career should finish on the same make on which he started back in 1934.

Jack Brett (1951 &1953-1955)

Jack Brett

1956 Senior TT

Yorkshire's Jack Brett was a comparatively late starter in the racing game. He was born in Leeds in 1918, younger by six years than his brother Charlie. The elder Brett made his Isle of Man debut in the 1935 Lightweight Manx Grand Prix before he moved to the TT. Thereafter his mounts were Excelsiors, Velocettes and Nortons until he left the racing to brother Jack. Following active service in the artillery, Jack Brett ended his war as a warrant officer in the RASC. No doubt influenced by Charlie, Jack's racing career began at the 1946 Manx Grand Prix.

He rode a 350 Velocette in the rain-deluged Senior and did well to gain a replica with his 7th place. He switched to the TT in 1947 and was holding fifth place in the Junior when his Velocette expired at Sulby. The following year C. Brett and J. Brett were starters in the Junior TT, each on Velocettes. Both retired, but in the Senior TT Jack's Norton finished in 8th place ahead of Italian ace Tenni, who made fastest lap on a Moto Guzzi before he was slowed by sundry problems.

For 1949 it was a Norton in both races for Jack. He finished well down in the field but at least he went the distance. The situation improved considerably in 1950; riding Nortons entered by Hallen Motor Engineers. A 15th place in the Junior TT and 8th in the Senior resulted in two silver replicas for a delighted Jack Brett. His inclusion in the 1951 Norton team came about through Dickie Dale's illness just before the TT. He justified the appointment with a fine 3rd place in the Junior. Jack's Senior race came to an abrupt conclusion when he dropped his Norton at the Gooseneck. A full round of the World Championship as a Norton teamster followed the TT; with Jack achieving leader board finishes in several races, and 5th place in the 350 World Championship.

At the end of the season he made the change to AJS, which appeared to be a propitious move when on his first outing he won the 350 class at the September Scarborough meeting. The opening classic of 1952 was the Swiss grand Prix at the leafy Bremgarten circuit. Jack's elation at winning the 500 race was saddened by Dave Bennett's fatal crash in the closing stages.

Brett's Isle of Man TT bordered on disastrous. His Junior AJS broke down on the Mountain and in the Senior he virtually destroyed the Porcupine twin in a crash at Quarry bends. He fared little better for the rest of the season, apart from a 4th place at the 350 Italian Grand Prix and 7th in the Senior. Jack's tenure with AJS had been less than memorable and he badly wanted a return to Nortons.

He was signed for 1953, with the proviso that he cut down on his gregarious life style. He did not finish on the leader board in the Junior TT, won by Ray Amm, and in a dramatic Senior it could well have been Jack's moment of glory, for he came tantalizingly close to winning the Senior TT. On lap 6 the order was Amm (Norton), Armstrong (Gilera), and Brett (Norton). On the final lap Amm had a minor spill and restarted, minus a footrest. Armstrong was delayed at Ramsay replacing the Gilera's chain. That turn of events could well have given Jack the Senior TT. He crossed the line first and then had to wait a frustrating period for Amm to finish. Finally Amm was flagged in with 12 seconds to spare and be declared the winner from a disappointed Brett. Armstrong finished third.

Norton's works team for 1954 featured Ray Amm again as No1, supported by Jack Brett and newcomer Bob Keeler. In 1953 Keeler gave Norton their 6th and last victory in the Isle of Man Clubman races. The 350 class was already dominated by the BSA Gold Star and from 1954 until the series was terminated in 1956 the 500 Gold Star achieved equal dominance.

Amm retired with mechanical problems in the Junior TT, giving AJS a 1-2 finish with Rod Coleman and Derek Farrant and in his first TT appearance Bob Keeler was third. In the controversial Senior TT the ever–reliable Jack Brett was third behind Ray Amm and Geoff Duke on the Gilera. Norton's racing policy underwent radical changes from 1955 to 1957. Rather than field pure works bikes it used development versions of the standard Manx.

The Slazenger organization provided finance for the team which comprised Jack Brett, John Surtees and John Hartle. Jack gave them a taste of victory in winning the 500 class of the 1957 North West 200 prior to the TT. More success came at the 1957 500 Belgian Grand Prix with Jack declared winner, following the disqualification of the original winner Liberati on the Gilera.

Brett's win should not be regarded as a hollow victory, for his race average was 113.35mph compared to Surtees' 1956 speed of 114 on the MV four. It was a race of high attrition with just five finishers; and with the first four riding Nortons, ahead of a lone BMW. (Eventually Liberati was re-instated as the winner)

By that stage classic race wins for Norton were the exception rather than the rule. Jack's last TT appearances were the 1960 Junior and Senior. He finished 17th and 23rd respectively, not the best note to finish on but he had enjoyed a great innings. Jack opted for a quieter life style on the golf course rather than the cut-and-thrust of racing. In 1982 he suffered a fatal heart attack while playing golf. With his loss the racing world was robbed of one its great characters.

R.H. (Dickie) Dale (1950-1951)

R.H. (Dickie) Dale

Brands Hatch 1959 (BMW)

Lincolnshire rider Dickie Dale was remembered more for his association with Italian factories. He raced a wide variety of makes and performed well on them all. His slight build and quiet manner belied a determined attitude to racing. He had a smooth sensitive riding style which suited him well to the ultra-light Moto Guzzis of that era. However it was on works Nortons that he was at first expected to leave his impression on the racing scene.

He was born in 1927 in Wyberton where his father owned a haulage business. From an early age his sights were set on a flying career. In 1945 Dale began his compulsory National Service with the RAF and instead of achieving his goal as a pilot he served as an engine fitter. Off-duty periods were spent racing a 350 Velocette at local grass track venues Dale was de-mobbed in 1948 and furthered his racing career at English road circuits, using a 500 Norton and 350 Velocette. His Isle of Man debut was the 1948 Lightweight Manx Grand Prix riding a 250 Moto Guzzi. He demonstrated his potential by winning the race in convincing style. More success came on the Guzzi at British mainland circuits, plus some good results on the Velocette.

At the 1949 Isle of Man TT he was provided with a works Guzzi for the Lightweight event. Dale climbed from 4th on lap one to 1st on lap five, smashing the lap record in the process. He was holding a three-minute lead ahead of the other works runners until the Guzzi succumbed to its Achilles heel, a broken valve spring. Dale's ability attracted the notice of Joe Craig who provided a works Norton for the 350 Ulster Grand Prix. He rewarded Craig's judgment with 6th place and was the first Norton home behind a gaggle of Velos and Ajays.

For the 1950 season he rode a privately entered 350 AJS and 500 Norton. At the TT he finished in 7th place in both the Junior and Senior events; first private entrant behind the works runners. Dale was back in the Norton team for the Dutch TT, finishing 7th in the 350 race. At the very wet Swiss Grand Prix he showed his versatility in taking a works Benelli into 3rd place in the Lightweight. Back in the Norton camp for the 1950 Ulster, Dale rode a steady race to finish 4th in the 500 event which was won at record speed by Geoff Duke.

The 1951 season began well for Dale with a win at the North West 200 in the 350 and 2nd in the 500 behind Johnny Lockett. More success came his way at the Eppynt circuit in Wales, with wins in the 350 and 500 finals riding the same 350.

These successes ensured him a place in the 1951 Norton team but he was to be cruelly denied his opportunity. Practice for the Isle of Man TT had barely started before Dickie was taken ill. His problem was first thought to be pleurisy but it proved to be far more serious. He was stricken with tuberculosis.

His convalescence was a prolonged and frustrating period and in no way did it blur his image as a sound team man. In 1953 he joined Duke and Armstrong at Gilera. On his first outing on the Italian four he scored a win at the North West 200 in the 500 race ahead of Armstrong. Four Gileras started in the Isle of Man Senior TT and just one finished; with Reg Armstrong in 3rd place. Duke went on to claim the World Championship with victories at the Dutch, Italian and Swiss Grands Prix. Dale got among the points at Belgium, Italy and Spain. His best result was 2nd place at Monza, behind Duke and ahead of teammate Liberati.

For 1954 Dale became a member of the rival MV concern. His best result came at the end of the season with

1st place at the Spanish Grand Prix. He also finished 4th at the 500c Ulster Grand Prix although his tenure with MV lasted only one season. In 1955 he began a long and fruitful association with Moto Guzzi where he was regarded as the perfect team man; utterly dependable and always prepared to do his best. A highlight of his initial season was first place in the 350 Italian Grand Prix.

At season's end, Dale and teammate Bill Lomas journeyed to Australia. The plan was to campaign a brace of Guzzis on circuits in New South Wales and Victoria and also enjoy a break from the European winter. They did manage to score wins at some venues against the locals.

Like previous overseas riders they too were surprised by the performance of the local machinery.

Lomas, like Dale was an experienced, analytical rider. He won the 1955 and 1956 350 World Championships on a Moto Guzzi and only for a technicality he would have gained the 1955 250 crown for MV as well. Dale gave the fabulous V8 its only Isle of Man outing with 4th place at the 1957 Golden Jubilee TT. At the close of the season, Moto Guzzi, Gilera and others pulled out of international racing. Without the security of a works ride many riders were facing redundancy. Duke and Dale equipped themselves with works-prepared BMWs. The flat twins from Munich dominated sidecar racing for a decade and apart from some fine performances by Walter Zeller they never found favour as a solo. Duke never came to terms with the BMW and reverted to the more amenable Norton. The BMW was a machine whose characteristics demanded a more physical type of rider. Dale persevered for two seasons before he followed a similar path to Duke.

He became one of the many private entrants who made up the field against the all-conquering MVs. In April 1961 the battling privateer was leading the field on his Norton in a 500 cc event at the notoriously difficult Nurburgring. Dale crashed heavily in the wet and treacherous conditions. He was critically hurt and sadly he succumbed to his injuries while being airlifted to a Bonn hospital. The Nurburgring was a tragic end to Dale's career, one that he pursued with modesty and dedication.

Harold Daniell (1947-1950)

Harold Daniell

1949 Senior TT

Londoner Harold Daniell was one of many top-flight riders whose careers were interrupted by World War 2. When racing resumed in 1947 he was able to continue as a Norton teamster. His teammates, Artie Bell, Ken Bills and Ernie Lyons all had pre-war experience although it was Bell who presented the greatest challenge to team leader Daniell.

In his very first Senior TT, Bell took the lead halfway through the race. It was a remarkable effort for a newcomer to the Island but at his pit stop Bell was ordered to ease his pace. Daniell went on to win his second Senior TT with Bell in second place. In third, and sharing fastest lap with Artie Bell was Peter Goodman; entered on one of the rare pre-war Velocette works 500s.

Daniell and Bell were the winners of most of the early post-war Continental Grands Prix. Harold enjoyed particular success at the Swiss Grand Prix, pre-war and post-war but one event was an ongoing nemesis for the bespectacled Londoner. He contested the Ulster Grand Prix from the mid-30s until his retirement from racing in 1950 and every year he was sidelined with either mechanical problems or illness. Finally, his personal gremlins relented for his final work's appearance, the 1950 Junior Ulster.

Throughout 1948 and 1949 Norton persisted with their outdated works machines; unwilling to follow the Italian trend of multi-cylinder design It was the introduction of the Rex McCandless frame that enabled Norton to remain competitive for years beyond what should have been a total eclipse. It was Harold Daniell who actually coined the phrase 'Featherbed' for the new frame, an appellation that endures to this day.

For the 1949 season Norton did not support a works team. Instead they loaned their 1948 machines to a syndicate composed of Daniell, Bell and Lockett with Steve Lancefield supervising their preparation. Their first outing was the North West 200 in Northern Ireland where Bell won the Senior, while Daniell scored a narrow win over Frith's Velocette in the 350 event.

Next came the Isle of Man TT series. Norton's biggest challenge was expected to come from the AJS works team. Their 'Porcupine' twin was originally designed as a supercharged unit and with the post-war ban on superchargers the design was severely handicapped. After two seasons that brought indifferent results, by 1949 the 'Porc' had achieved a degree of reliability and speed to make it a genuine contender for Grand Prix success.

However it was Bob Foster on the lone Moto Guzzi V-twin who startled both Norton and AJS camps and seemed poised for a Senior victory. On lap one he was fourth behind the AJS pair of Graham and Frend and Daniell's Norton. Lap two saw a three-way tie for the lead between the Ajay duo and the flying Guzzi. On lap three Foster went into a straight lead, which he maintained on lap four and improved on it by lap five.

The race looked a certainty for Foster, sadly it was not to be. Transmission problems caused a retirement at Sulby, giving the lead to Graham with Daniell inheriting second place, Harold was experiencing severe vibration problems that brought him to a temporary halt on the final lap. He restarted and continued to the finish line, unaware that Graham had broken down at Hillberry and was pushing his machine to the finish. Thus, Harold Daniell won his third Senior TT with Lockett second and Bell fourth. Although outpaced by its rivals the somewhat dated Norton had outlasted them to score another Isle of Man victory.

In 1950 the Norton ream was revitalized with the introduction of the 'featherbed' frame and the signing of Geoff Duke. Their TT results were reminiscent of Norton's glory days of the early 1930s, with a Bell, Duke, Daniell 1-2-3 in the Junior and a Duke, Bell, Lockett 1-2-3 in the Senior. Without doubt their rising star was Lancashire's Geoff Duke. Only for tyre failure which eliminated the entire Norton team in the Belgian GP and Dutch TT, Duke would have certainly become 500cc World Champion in his first Grand Prix season. It has been suggested that Daniell never quite came to terms with the 'Featherbed' or the phenomenon that was Geoff Duke. One suspects that he considered himself equal in ability to Bell and superior to Lockett but this new arrival simply outclassed everyone.

Daniell's impressive record spanned two decades and apart from three forgettable seasons with AJS it was exclusively on Nortons. His final Grand Prix placing was 6th in the 1950 Junior Ulster Grand Prix; a modest result although it did represent a finish at a venue where good fortune had rarely gone his way.

Following his retirement Daniell was able to direct all his energies to his South London motorcycle dealership. He tended to specialize in the sale of race bikes, and not surprisingly these included Manx Nortons. Aside from that Harold found time to campaign a Formula Junior race car in the mid- 1950s, which proved to be only a brief diversion for the burly Londoner. Failing health caused him to retire from racing and shortly afterwards the Norton campaigner and triple TT winner passed away.

Geoff. Duke (1950-1952)

Geoff. Duke

1951 Junior TT

How would it be possible to do justice to the career of
Lancashire's Geoff Duke within the strictures of this
volume? There can be no doubt that he was the most
outstanding rider of his era...'of all time', many would insist
and it is a difficult claim to refute. That he became involved
in racing was remarkable for he was raised in a family
situation that was implacably opposed to motorcycles.

Geoffrey Duke was born at St. Helens in 1923. His first
employment situation was as a technician in the postal
service. From there he enlisted in the Royal Corps of
Signals and attained the rank of sergeant. During his army
tenure Duke was an enthusiastic member of the signal
corps motorcycle display team. Following his de-mob in
1947 he purchased a 350cc competition BSA, and on the
'Beeza' he began his career as a trials rider.

Duke's exploits attracted the attention of Artie Bell who
was instrumental in finding a position for him at Norton. At
first he had scant success on the 500T trials Norton, which
was quite a deal heavier than the BSA. The prospect of
becoming involved in road racing seemed much more
appealing. Nortons supported him fully from the outset; to
the extent of loaning him a 350 Manx for the 1948 Manx
Grand Prix. Earlier he had applied for an entry in the 1948
Isle of Man Clubmen's TT but the series was
oversubscribed and he was rejected. His next option was
the Manx Grand Prix which would be his first ever road

race. He arrived in the Island a week before official practice began and using his trials machine, fitted with road tyres he set about learning that most difficult course.

He did not feature on the practice leader board. His practice plan was to conserve his machine on the straights and concentrate on the quickest possible way through the numerous bends. On race day he faced the starter; quietly confident of a good showing. Race favourite was veteran Denis Parkinson, riding a Francis Beart-prepared Norton. Parkinson was already a triple Lightweight winner at pre-war Manx Grands Prix. Duke ignored such reputations and rode to a pre-determined plan that saw him assume the lead on lap 3. Shortly afterwards he was forced to retire with a split oil tank, which was a great disappointment and only for his problem he may well have won a Manx Grand Prix at his first attempt.

Duke's first Isle of Man success came at the 1949 Senior Clubman's TT, in which he was entered on a Norton International; a roadster version of the Manx. He dominated the race from start to finish and in the process he set a lap record that stood until 1953. Geoff was anxious to make a good showing at the 1949 Manx Grand Prix in September but in July he was involved in a crash at the Skerries 100 in Ireland. He suffered a broken leg and painful lacerations when he was thrown through a thicket hedge and had barely recovered in time when practice began for the Manx.

His main opposition at the Manx came from Ulsterman Cromie McCandless, entered on Beart-prepared Nortons. McCandless was the younger brother of engineer Rex and like Duke he was a most capable rider. Both riders encountered problems during their races which hampered their prospects of winning. Duke was lying a close second to McCandless in the Junior event and was forced to lay down his Norton at Ramsey Hairpin to avoid a collision with a slower rider. Duke managed to restart his damaged machine and pressed on to still finish in second place. Positions were reversed in the Senior Grand Prix with Duke consolidating his lead ahead of McCandless who lost time on the final lap with fuel problems.

Following his Manx success Duke was an obvious selection for the 1950 works team. His initial outing was in late 1949 at the Montlhery speed bowl near Paris for the purpose of attacking some current world records. Artie Bell was Duke's partner in the attempts which resulted in a bag of 21 records. To be included in the Norton team on equal terms with Bell, Daniell and Lockett was really special and to start the 1950 season on the new-look 'Featherbed' was an opportunity beyond compare. Duke scored emphatic wins in the 500 class of the Senior TT, the Ulster Grand Prix and the final Grand Prix of the season, the Italian.

Only for tyre failure in Belgium and Holland he would have become 500cc World Champion at his first attempt. The following season he was doubly vindicated, becoming World Champion in both 350 and 500cc categories. This was a remarkable achievement; in particular to claim the 500 crown; riding a machine that was clearly slower than the Italian multis which were his main opposition. Duke remained loyal to Norton for the 1952 season and in spite of his undoubted ability it was obvious he was fighting a losing battle. He did gain the 350 World Title but by that stage the 500 Norton was quite outclassed.

Interspersed with motorcycle racing Duke made a foray into driving for Aston Martin in sports car races. It was an unhappy episode in his career, due mainly to resentment from other drivers, in particular his teammate Peter Collins. He did enjoy reasonable success at mainland events during 1952 but his tenure with Aston Martin came to an abrupt end at the 1953 Sebring 12 Hours Race held in Florida. Collins had put the Aston into a comfortable lead at the time of his first pit stop and driver change. Duke took over, determined to maintain the lead and during an overtaking manoeuvre he spun out and damaged the car too seriously to continue. Collins was most scathing about the incident which upset Duke greatly. The whole affair convinced Duke that his real future was on two wheels.

Apart from that disappointing outcome, Duke was bewildered by Norton's attitude regarding his future prospects with them. In his three seasons with Norton he had ridden brilliantly to gain 3 World Championships for the firm. Then at the start of the 1953 season Nortons announced to the Press; "There is no place for Duke in the Norton team." Later, Duke discovered that the objection to his re-joining the team came from his old teammate Ken Kavanagh.

If Norton did not want Duke's services, then Gilera were more than anxious to sign him and with a retainer and benefits that made Norton's salary look decidedly paltry. It was the commencement of a 5-year tenure with the Italian firm, which gained the St. Helens star a further 3 World Championships, (1953-54-55) including the 1955 Senior TT at record speed.

At the end of the 1957 season the leading Italian factories withdrew from Grand Prix racing. Without a regular works ride, Duke campaigned a factory-prepared BMW and apart from a win on the BMW at Hockenheim he had scant reward for his efforts. He enjoyed more success with a brace of standard Manx Nortons, in particular at the Hedemora circuit in Sweden where a scored a hard-fought Senior/Junior double.

Duke's final season was in 1959 riding Manx Nortons. This was his Isle of Man swansong where he gained a doughty 4th place in that year's Junior TT. His last race wins were in September at the non-championship Swiss Grand Prix at Locarno where he scored a Junior/Senior double, plus a Lightweight win on a 250 works Benelli.

Following the Locarno meeting he announced his retirement from motorcycle racing. In 1960 he was given the opportunity to resume a racing career on four wheels. An organization named Chequered Flag invited Duke to drive for them in Formula Junior events. He competed in several races but the operation was under-funded and Chequered Flag withdrew from racing. This was a disappointing outcome and Duke settled back into personal business commitments.

Then out of the blue, in June 1961 he received an offer to drive a Formula 1 Cooper at a meeting at Karlskoga in Sweden. He was not impressed with the circuit and even less with the car which gave trouble during practice. Against his better judgment he started in the race and on the fourth lap the gearbox locked solid. The car hit a grass bank and overturned, flinging out the luckless driver. Duke landed face down and was getting to his feet when the car landed back on top of him. He suffered extensive injuries, far more serious than he at first realized. At that point he retired from competitive motoring.

The 'Duke' still made parade appearances on two wheels where a younger generation had the opportunity to witness one of the truly 'greats' in action. His name is perpetuated in a range of motoring videos marketed by his son Peter.

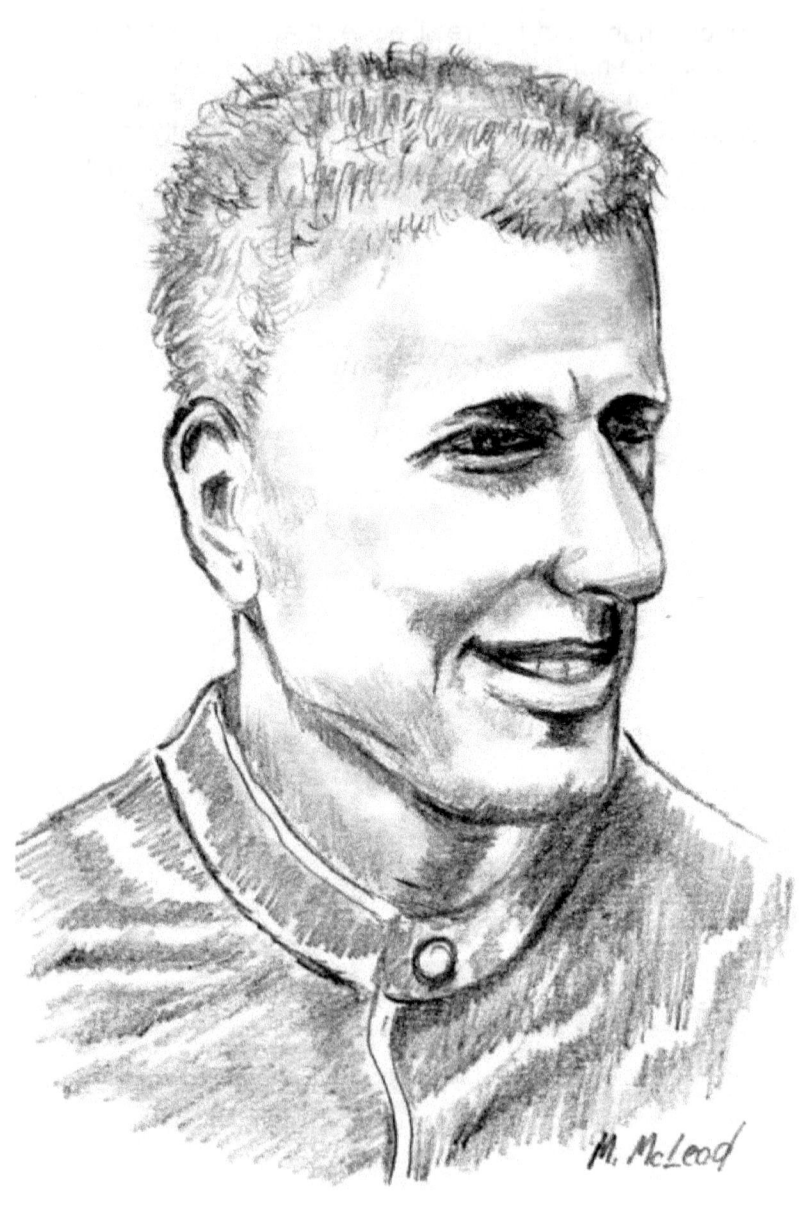

John Hartle (1955)

John Hartle

1957 Senior TT

Derbyshire's John Hartle is remembered as the 'almost' man of motorcycle racing. He was a fast, stylish rider who never seemed to take ridiculous risks, yet he was involved in serious crashes that be-devilled his career. Several big wins came his way during a career that began at the Oliver's Mount circuit at Scarborough in 1954. The supreme irony was that it should end fatally at the same venue in 1968 when he collided with the framework of an overhead pedestrian footbridge.

Geoff Duke had the opportunity to assess Hartle's ability during the 1963 season when he launched his bid for a Gilera comeback. He engaged Derek Minter as No1 for his 2-man team, backed up by Hartle. Duke was confident of Hartle's skills at the Isle of Man but held doubts about his ability to sustain the pressure of the cut-and-thrust of Grand Prix racing. His assessment was that Hartle was good though not superb and lacked the urge to ride ultra-hard to win.

Be that as it may, Hartle displayed great potential at the 1954 Senior Manx Grand Prix, which was only his second essay at the Isle of Man and during a rain-soaked race he led into the final lap, only to be robbed of victory when his Norton ran out of fuel. His third place on an AJS in the Junior Manx was of some consolation although nothing matched a win in the Senior.

In 1955 Hartle became a member of the Norton works team, partnering John Surtees and Jack Brett. He secured 6th place in the Junior and 13th in the Senior, and in the following year Hartle was in the Slazenger-sponsored Norton team. He did well to finish 3rd in the Junior TT, behind Ken Kavanagh's Moto Guzzi and Derek Ennett's AJS. In the Senior he finished second to his former Norton teammate John Surtees, in his first season with MV Agusta. Derek Ennett was poised for a successful racing career. He won the 1953 Junior Manx on an AJS; possibly with works support and following his good result in the 1956 Junior TT he gained a works Moto Guzzi for that year's Junior Ulster Grand Prix.

Still coming to terms with the unfamiliar Guzzi he crashed and was fatally injured. The Junior Ulster saw Lomas and Dale on Moto Guzzis finish 1st and 2nd ahead of Hartle's Norton. It was a creditable finish for Hartle and in the Senior Ulster he outlasted the strong Italian challenge to finish in first place.

In 1958, with Surtees' influence, Hartle became a team member with MV and rode the Italian fours in the Island in 1958 and 1959. He suffered 3 retirements from 4 starts in that period and did finish second to Surtees in the 1959 Junior. His undoubted ability on the MV was demonstrated in 1960 when he finished second to Surtees in the Senior. Earlier in the week he scored a popular win in the Junior ahead of Surtees' ailing MV, with Bob McIntyre (AJS) in 3rd place.

Riding the Gilera in 1963, Hartle was overshadowed at the TT and other classics by an in-form Mike Hailwood. At the Dutch TT Hailwood experienced mechanical problems, allowing Hartle to win and give Gilera a brief moment of glory. The attempted Gilera comeback was a bold venture which cost Duke a considerable amount of money and a great deal of heartache. It unfortunately coincided with Hailwood at the top of his form; although a major factor was Minter's serious crash early in the season. He suffered a broken back in a horror crash at Brands Hatch that put him out of racing for most of the year and he only returned for the last of the Grand Prix events. In Duke's estimation Minter was the only rider capable of matching Hailwood.

Hartle experienced a bout of serious crashes over the following seasons that were enough to sideline him until 1967. After such a prolonged lay-off he made a somewhat reluctant return to racing. He scored wins at Oulton Park and in the 750 Production TT at the Isle of Man, all of which were a great boost to his confidence. Riding a G50 Matchless he contested the classic events and finished the season with second place in the 500cc World Championship.

He was looking forward to an even better season in 1968, particularly at the TT where he secured MV rides in the Junior and Senior. The previous year Hartle scored a convincing win in the 750 Production TT on a works Triumph Bonneville and was keen to repeat the performance in the 1968 race. He was possibly trying too hard in the wet conditions, crashed heavily and was pronounced unfit to start in the Junior TT.

He was well enough to start in Friday's Senior, yet again he was thwarted when the MV developed gear election problems that worsened during the opening lap. His battle with the recalcitrant machine terminated at Cronk-ny-Mona when the MV went into a vicious tank-slapper, flinging Hartle off at high speed. Any hopes of further MV rides evaporated in the crash and his initial reaction was to retire from racing immediately and permanently. After deep soul-searching he decided otherwise. That indomitable spirit which enabled Hartle to shrug off serious injuries also caused him to ignore any signs of impending tragedy and three months later he was claimed by his nemesis, Scarborough. It was tragic that Hartle never won a World Championship, only for machine breakdowns and crashes that crown would surely have been his.

Harry Hinton (1949-51)

Harry Hinton

1951 Junior TT

Harry Hinton Senior was patriarch of a family dynasty that encompassed three generations of motorcycle racers. He was born in Birmingham in 1911 and immigrated to Australia with his parents in 1920. Perhaps there was something prophetic about his birthplace, for it was also the home of Norton and BSA, two makes with which he was associated for much of his career.

He grew up in Sydney's western suburbs and his early forays were beach races at Gerringong, south of Sydney. In 1931 he lost an eye in a serious road accident that could well have ended his racing career. It took time to adjust to the loss and it was a tribute to his determination that he returned to racing.

In 1932 Harry acquired a Norton International; a potent machine which provided him with some notable wins. Already he had a reputation for making unfancied bikes competitive and competitive bikes even faster. An opportunity to demonstrate his talents came in 1933, with his appointment as trade rider/tuner for BSA agents Bennett & Wood. Following their disastrous foray at the 1922 Isle of Man TT, BSA were opposed to building outright racers and sports models. Under Harry's influence the rather pedestrian Blue Star range became competitive enough to challenge the race-bred camshaft Norton, AJS and Velocette machines.

The BSA lacked stamina when ridden hard on long straights but they were ideal on short circuits where their acceleration matched the more sophisticated 'cammys'. Harry Hinton and his flying BSAs became household names during the 1930s; and to demonstrate his versatility he would win a race on the 500 and then bolt on a sidecar and continue his winning ways.

Post-war he concentrated on Nortons in the larger capacity classes, with one notable exception, the 1948 Australian Senior TT at Mount Panorama. On that occasion Harry rode one of the new Grand Prix Triumphs rather than his usual Norton. Victorian rider, Frank Mussett represented his biggest challenge, riding a rare ex-works 500 Velocette, which proved to be a comprehensive victory for Mussett and in the process he created a lap record that stood until 1951.

In 1949 Harry Hinton and Eric McPherson were selected by the ACCA as official Isle of Man representatives. A third rider, George Morrison from Ballarat accompanied them but as a private entrant. Before leaving for England Hinton rode at the Easter Bathurst meeting. He won the 250 class on a borrowed MOV Velocette, instead of his usual BSA and completed a hat trick of 350 wins on his Norton, following his successes in 1947 and 1948.

Hinton and Morrison made impressive debuts at the TT; with Hinton finishing 15th in the Junior and a creditable 9th in the Senior. Morrison looked set for an even better finish in the Senior but fell victim to the Norton bogey of a broken frame. He pushed and coasted in to finish 31st. The Aussie pair rode in the classic Grands Prix and also at minor venues where both scored wins. As a reward for their efforts both riders were provided with works Nortons for the Ulster Grand Prix.

At season's end Harry Hinton returned home and included in his luggage was a piece of Norton history; Artie Bell's 1948 Senior TT winner. Harry used it to good effect in winning the 1950 Australian Senior TT at Ballarat in January. Before he left for overseas he rode at the Easter

Bathurst meeting where he finally scored that elusive Senior TT win.

Hinton, Morrison and McPherson were selected as Australia's 1950 Isle of Man team, with Hinton and Morrison drafted into Norton's 'B' team, alongside veteran Harold Daniell. Their 'A' team comprised Duke, Bell and Lockett, all on the new 'Featherbed' model. If a spare became available it was allocated to Daniell. Hinton finished in 10th place in both Junior and Senior TTs, riding the outmoded 'Garden gate' model. Morrison had mixed fortunes in his races; retiring in the Junior and finishing 11th in the Senior.

Following the TT, the Australians travelled the continent, riding in the classic Grands Prix and minor events. The Belgian Grand Prix was a memorable one for Norton and for the very worst reasons. First there was Artie Bell's horrendous crash, followed by tyre problems that eliminated the entire Norton team. Hinton's was the first Norton to finish; his 6th place gained him the first Championship points scored by an Australian in the 500 category.

The Dutch TT, held on the following weekend brought even more drama. Hinton finished 6th in the Junior behind 3 works Velocettes and 2 works Nortons and ahead of the entire AJS works team. Morrison crashed on the first lap and broke a leg. He never raced again in Europe. The Senior race was barely into its stride when the entire Norton and AJS teams, either crashed or withdrew with tyre problems. It fell to Hinton to uphold British prestige and in the ride of his career he took on the leading Gileras, to eventually displace one of them and finish a heroic third.

He was recalled to the works team for the Ulster Grand Prix and finished third in the Junior behind the Velocettes of Bob Foster and Reg. Armstrong. The Italian Grand Prix at Monza was the final event in a tumultuous season. Retained once again by Norton, Hinton rode a fine race to finish third in the Junior, backing up team leader, Duke. In so doing he deliberately sacrificed his second place to Les Graham's AJS in allowing Duke to win.

After his fine showing during the year Hinton was assured of a place in the 1951 works team. He won the 500 and Unlimited events at the Easter Bathurst meeting and then sailed for England. TT practice week saw him consistently on the leader board and his lap times on a 250 Moto Guzzi were quite amazing.

Maurice Cann suffered a practice crash and was unable to race his famous 250 Guzzi. He sportingly offered the ride to Hinton. Harry soon adjusted to an unfamiliar machine and in a spectacular practice session he slashed 60 seconds off the existing lap record. He went into race week as clear favourite for the Lightweight TT and a strong contender in the Junior and Senior. He began the Junior TT in great style, holding second place behind a flying Geoff Duke and ahead of Norton teamster Lockett. Disaster struck on lap three when Hinton crashed heavily at Laurel Bank and sustained multiple breaks in one hand, plus a splintered kneecap. He never raced in Europe again. Meanwhile his Australian career went from strength to strength, particularly at Bathurst.

In 1952 he scored a Junior/Senior double and in 1953 he made it a quadruple, winning the Lightweight, Junior, Senior and Unlimited events. The following year was unproductive, for he crashed in practice and broke a collarbone. In Hinton's absence the versatile Kiwi, Rod Coleman scored his own Senior/Junior double. Harry's celebrated Bathurst career ended in 1955 with another Junior/Senior double and at Mount Druitt in July he made his final race appearance. He passed his machines on to his eldest sons Harry and Eric who were already making an impression on the local racing scene. Harry Hinton died in 1978, aged 67; a highly respected member of the local and international racing fraternity.

Ken Kavanagh (1951-53)

Ken Kavanagh

1952 Senior TT

Ken Kavanagh's racing forays were typical of the Australian riders who made the pilgrimage to Europe in the early 1950s. Veterans, Harry Hinton and Eric McPherson who preceded him both earned World Championship points and also secured works rides. Kavanagh was determined to make his mark in Grand Prix racing and succeeded far beyond his contemporaries. During his career he gained works rides from four manufacturers; Norton, Moto Guzzi, MV and Ducati.

Kavanagh was the first Australian to win a classic Grand Prix and to also win at the Isle of Man; in his case, the 1953 Senior Ulster and the 1956 Junior TT.

His competition career began in motocross or scrambles as it was known in Australia. He moved from there to road racing, riding a new Manx Norton; the first to be imported post war to Australia. Kavanagh became the man to beat in Victorian road races. His big opportunity came in 1951 when he was chosen for the Isle of Man team, partnering Harry Hinton and Tony McAlpine.

The TT was Kavanagh's main focus on his arrival in Europe. He applied himself to learning that complex 37-mile circuit in a disciplined manner, with fast lap times and without the ignominy of crashing. A feature of Kavanagh's career was his ability to avoid dropping the model. These were times when many of the circuits were incredibly dangerous, with no run-off areas and road conditions that would be quite unacceptable today. Under such circumstances there were inevitable fatalities, including many aspiring Australian riders.

Kavanagh was actually offered a works Norton for the Senior TT. Hinton's injury in the Junior TT left the team one rider short and it was a rare compliment for Kavanagh, on his first Island foray. He rejected the offer, saying that his own Manx Nortons were quick enough for his current experience. During the Senior TT he worked his way into fourth place, only to retire with a split oil tank.

Following the TT Kavanagh scored his first ever-European victory with a win in a 500cc race at the Thruxton circuit. He was also successful at the Tarare circuit in France, winning the 350 and 500 races. Other wins followed his French outings and for the Ulster Grand Prix he was enlisted into the Norton team alongside Geoff Duke, Johnnie Lockett and Jack Brett. Kavanagh justified his inclusion with a strong second place to Duke in both the 350 and 500cc events.

Kavanagh returned to Australia in October, following a superb performance by a newcomer to European racing. Back in England for the 1952 season he scored 350 and 500cc wins at Boreham Wood, making him a British Champion. His Isle of Man Senior outing, like the previous year proved disappointing. He was holding fourth place on the final lap when his primary chain broke. Determined to finish he pushed and coasted the final six miles to come 32nd. His Junior TT resulted in retirement with mechanical problems, again while holding 4th place.

It was a season of mixed fortunes for the 1952 Norton team. At the German Grand prix on the Solitude road circuit Ray Amm crashed in practice, sustaining a broken leg and other injuries. Team leader Duke was also absent, following a crash at Schotten on the previous weekend. It was left to Kavanagh and Armstrong to uphold Norton's honour. Kavanagh was poised to win the 350 race only to have Armstrong beat him to the line by a wheel's length.

His Senior race was an example of riding to team orders. In his efforts to keep ahead of Graham's MV, Kavanagh built up an adequate lead over the MV and Armstrong's Norton. Graham encountered problems, putting Armstrong into second place. At the time Armstrong held a slender lead in the 500cc World Championship, making it imperative that Kavanagh surrender his first place.

He duly slowed his pace to allow Armstrong to come through for his second win of the day. That year's Ulster Grand Prix brought a more satisfying result for the Australian with first place ahead of Armstrong in the 350 event. It was the final year that the GP was held on the old Clady circuit, prior to its move to Dundrod. Armstrong needed a win in the 500 race to consolidate his narrow points lead in the World Championship but he was forced to retire with a broken chain. It was an exasperating occurrence that bedevilled Nortons on so many occasions. Cromie McCandless won the Senior, riding a work's Gilera.

For 1953 Kavanagh was once more a Norton teamster alongside Ray Amm and Jack Brett. Geoff Duke and Reg Armstrong had moved to Gilera, following an unprofitable season for them on Nortons. Despite the speed advantage of the Italian fours the Norton's superior handling kept them competitive at tortuous circuits like the Isle of Man. Kavanagh was confident of a good result at the Junior TT and came tantalizingly close to achieving that win.

On lap four he assumed the lead, ousting early leader Ray Amm. The Rhodesian was in tigerish mood and not to be denied. By dint of a record-breaking final lap he came through to beat Kavanagh by the margin of just nine seconds! Friday's Senior TT was even more dramatic for Amm. Following Les Graham's fatal crash on lap 2 Amm held second place behind Duke's Gilera

Then on lap 3 Duke came off at Quarter Bridge, damaging the Gilera too seriously to continue. Amm took over in first place, holding it till the finish with a record lap at 97.41 mph. Early in the race Kavanagh was well in the picture, climbing to fourth place by lap 3 only to retire shortly afterwards.

Duke's move to Gilera was clearly vindicated with wins at the 500cc Dutch TT and the Swiss and Italian Grands Prix. Kavanagh secured 4th place at the 500 Belgian GP, however the 500 Ulster Grand Prix was his moment of triumph. He was hoping for a leader board finish behind the Gileras until fate took a hand with torrential rain and clutch problems with Duke's Gilera. Midway through the race Kavanagh took over the lead while Duke was refuelling. Riding as never before in the rain-lashed conditions he held on to score a memorable win and become the first Australian to win a classic Grand Prix.

Kavanagh's final appearance on a work's Norton was an end-of-season meeting at Silverstone. It was a circuit that he disliked, due to its continuous use for car racing which left the surface absolutely without grip. He was untroubled to win the 350 race while the 500 final was a much more demanding exercise. Kavanagh was slipstreaming Dale's Gilera when he was brought down by the Gilera suddenly locking up. The Aussie was soon on his way again, regaining the lead to score his last win for the firm.

From 1954 to 1957 Kavanagh was retained as a works rider for Moto Guzzi, which was a productive era for the diminutive Australian with wins at the 1954 350 Belgian GP, the 1955 350 Dutch TT and a strong third place in the 1955 Senior TT behind Duke and Armstrong's Gileras. More significantly he won the 1956 Isle of Man Junior TT, becoming the first Australian to attain victory at the Isle of Man, a venue where good fortune rarely smiled on Thomas Kenrick Kavanagh.

His fellow riders were divided in their opinion of Kavanagh. Some considered his forceful ambition and total commitment as unsavoury. Without doubt he secured deals that made him by far the most successful of his contemporaries. Much of his wealth was lost in a foray into Formula One during 1958 and 1959. He was in a position to acquire a pair of ex-works lightweight Maserati 250Fs. His partner in the venture was to be fellow Aussie Keith Campbell, sadly the former 350cc world champion was killed at a minor race meeting at Cadours early in 1958. Motor racing soaks up a private entrant's funds at an alarming rate, as Kavanagh and many others have found, to their cost. After two unproductive seasons he abandoned the exercise.

In 1959 he made a return to two wheels with race entries on Nortons, the make on which his career began. He also campaigned Ducatis in the 125 and 250 categories. By then the Japanese influence was being extended into road racing, which was an invasion that Kavanagh resented. He had no interest in racing for them and on that note Kavanagh retired. He abandoned his Australian connections and settled in the northern Italian town of Bergamo. There he operated a dry cleaning business, a far cry from the glamour and excitement of his Grand Prix racing days.

R. St. John Lockett (1947-1951)

R. St. John Lockett

1949 Senior TT

Johnny Lockett was another rider whose career was interrupted by World War 2. He came into prominence with a fine second place in the 1938 Senior Manx Grand Prix, and only for the war he may well have advanced to the top rung of his profession. Throughout his career he was 100% Norton; as a private entrant in pre-war events and as a works rider from 1947 to 1951. His riding style was exceptionally neat but deceptively quick and it was a rare occurrence for him to drop the model. His pre-war rides were centred on Donington Park, the Isle of Man and the Brooklands speed bowl. Few riders were more consistent than Lockett at such diverse circuits as these.

His exploits attracted the attention of Brooklands identity, Francis Beart who was also a rider although his reputation was in race tuning. He became a legendary figure for his meticulous approach to machine preparation and from the mid-1930s until the 1960s he concentrated almost exclusively on Nortons. His list of successes was truly phenomenal, in particular at the Manx Grand Prix where from 1938 to 1975 his riders claimed 11 first places, 10 seconds and 5 thirds.

Beart's association with Lockett was curtailed by the war and in the two seasons preceding it they achieved some creditable results. Beart's initial Isle of Man foray was the 1938 Manx Grand Prix where Lockett was entered to ride his own 350 Norton. An unexpected bonus was the loan of a works 500 motor that could be fitted into the 350's frame and provide a mount for the Senior race.

The Manx was a distinct challenge for Beart, as it was his first experience of a long distance race, compared to the shorter mainland events. Lockett showed up well during official practice on the 500, with lap times within seconds of the record. On race day he made an excellent start and led for the first three laps. Then, he was forced to slow down due to a misfire and a split oil tank that was depositing oil on his rear tyre. Ken Bills on another Norton took over the lead and held it to the finish. Lockett managed to hold onto second place despite a challenge from Tommy McEwan, also on a Norton. The result was something of a disappointment for Lockett and Beart, but still commendable for a first attempt.

Lockett became well established as a rider; the names Beart and Lockett being almost synonymous. The 1939 International Grand Prix at Donington was an exceptional outing for the duo. Harold Daniell provided fierce opposition in the Junior and Senior events, although it was Lockett who prevailed in both races, and for good measure he also won the Unlimited event.

He was quietly confident of a good showing at the TT races which followed the Donington meeting. On several occasions he made an appearance on the practice leader board, and was clearly the quickest of the private entrants. He was beset by problems in the Junior TT; and after holding 9th place he experienced problems and eventually pushed in from Governor's Bridge to finish 16th; with some consolation in winning the Newcomer's Award and a Silver Replica.

In theory he had a better chance in the Senior TT; his machine having the work's motor, plus forks and wheels, supplied by Nortons. The Senior proved to be an anti-climax. He lay 7th at the end of lap one, dropped a place into 8th on laps 2, 3 and 4, and then, following a refuelling stop his motor expired at the 11th Milestone. It was a major disappointment, for he was again the leading private entrant. It would be nine years before he appeared at the TT again; and that was as an official Norton team member.

Lockett was sidelined by a motor accident shortly after the war and did not resume racing until 1947. His first works outing as a Norton team man was the Ulster Grand Prix. This was a gratifying outing for Norton; with Bell winning the Senior race and Lockett the Junior. His TT debut on works Nortons was the 1948 Junior, with Bell and Lockett finishing 3rd and 4th behind the flying Velocettes of Frith and Foster. Lockett's Senior race ended with a retirement on lap 2 after holding fourth place on the opening lap.

The Belgian Grand Prix followed the TT, and it saw total dominance by British machines. Bob Foster led a Velocette 1-2-3 in the Junior race ahead of Lockett's Norton, however the Senior was Lockett's finest hour. The opening laps saw a strong challenge from the AJS duo of Graham and West to the Norton trio of Bell, Lockett and Daniell. Also making a charge was a phalanx of GP Triumphs with Foster at their head. Bell encountered problems and dropped back, leaving Lockett equal to the situation and taking the lead which he held to the finishing line.

Perhaps the most memorable Grand Prix of the season was the Ulster; that year's Grand Prix of Europe; but for Norton it was for the wrong reasons. Irish weather can be most fickle and that year's Ulster was run in torrential rain and gale force winds. Graham, Bell and Lockett battled fiercely for the lead in the early stages and as the race progressed each rider experienced problems. Their misfortune became another's gain; with the Italian star Enrico Lorenzetti on a Moto Guzzi coming through to score a well judged victory.

Norton withdrew from supporting a works team for 1949 and instead loaned last year's models to a team composed of Bell, Daniell and Lockett. Apart from the TT series it was a fairly lean year for them, although Daniell outlasted his rivals to win his third Senior with Lockett 2nd and Bell 4th. The following year saw a marked change in Norton's fortunes, due to the revolutionary 'Featherbed' frame and the brilliance of Geoff Duke.

Artie Bell led a Norton 1-2-3 at the Junior TT, with Lockett in 6th place behind two works Ajays. There was a better result for him in the Senior TT, with a Duke, Bell, and Lockett 1-2-3. He did not feature strongly in the classic Grands Prix, apart from 4th place in the 350 Dutch TT and 3rd in the Senior Ulster.

For 1951 which was to be his final year, Lockett was teamed with Geoff Duke and Jack Brett with Ken Kavanagh joining them after the TT. Duke started the season in convincing style with a win in the Junior TT. In second place after a steady backing-up ride was Lockett, with Brett in third place. The Senior TT was almost a repeat of the Junior, with Lockett holding second for six laps behind Duke. Then on lap 7 Lockett retired with a broken chain; this was cruel misfortune after such a consistent ride. His retirement made way for the popular Bill Doran to claim second on the AJS Porcupine with Cromie McCandless achieving his best Senior TT result in third place.

For the remainder of the season Lockett featured among the place-getters, with 2nd in the 350 Belgian GP, 3rd in the 350 Ulster and 4th in the 350 French GP. At the end of the year he retired from motorcycle racing and directed his energies to motor racing. Entered on works MGs, Lockett partnered the Anglo-American driver Ken Miles on several occasions. They finished in 12th place in the 1955 Le Mans endurance race, which was remembered for the tragic accident involving Pierre Levegh's Mercedes. Throughout his motorcycle career Johnny Lockett was the epitome of dedication as a team member. The undoubted brilliance of Geoff Duke tended to overshadow his teammates and created a situation that relegated them to place-getters. Only for the Duke phenomenon, it is reasonable to assume that Lockett would have appeared more frequently on the winner's podium.

Ernie Lyons (1947)

Ernie Lyons

1947 TT Series

Ever-smiling Ernie Lyons was generally associated with the Triumph marque. Riding the Coventry twins he scored some notable wins in his native Ireland and in the Isle of Man. Velocettes also featured in his road racing career, particularly during the 1949 season when he partnered Freddie Frith on Velos sponsored by Nigel Spring. During the 1948 season he made appearances on Moto Guzzis and AJS 7Rs. His least successful season would have to be 1947, the year that he rode official Norton race ware. By all accounts it was a generally miserable period in the all-rounder's career.

Ernie's racing career began in 1932 while he was still at school. His first bike was a belt-drive 1912 Triumph which he campaigned despite its primitive design. He later changed to a 1927 AJS and began to make his mark in local races. Apart from road racing he was also a successful trials and scrambles rider.

Anxious to graduate to a more sophisticated machine for the 1938 season, Ernie's had a preference for a new 'cammy' AJS. With a price tag of £105, it was far beyond his resources. A better option was a new Triumph 'Speed Twin' that offered a similar performance and cost £75. Ernie was able to purchase one for £64. The package came as a collection of parts, minus electrics and mufflers, and it turned out to be quite a successful race bike. One notable result was 2nd place at the 1938 N/W 200.

The highlight of Ernie's season was the 1938 Senior Manx Grand Prix and his first essay at that challenging circuit. His Triumph performed well during practice and on race day he was hoping for at least a favourable finish. By all accounts it was a hectic race for Ernie Lyons. On several occasions he had minor skirmishes with corners and grass banks until he finally overdid things at the Gooseneck on his third lap and crashed out. Ernie was looking forward to a better showing at the 1939 Manx. Unfortunately it was cancelled due to World War 2.

It was a remarkable achievement for the Manx MCC to stage the 1946 series, following the conflict. Many of the pre-war riders were present, including the persistent Ernie Lyons. Again he was entered on a Triumph twin and on this occasion he did enjoy factory support, albeit in a clandestine manner. Its engine was unique in that it was an all-alloy unit, similar in design to those fitted as generators to RAF wartime bombers. Another feature was Triumph's patented sprung hub that served as an alternative form of rear suspension and a rather primitive one at that.

Ken Bills demonstrated that he had lost none of his pre-war brilliance with a win in the Junior Grand Prix. He was also a strong favourite for the Senior Grand Prix, a race that was made up almost entirely of Nortons. The weather forecast was for wet conditions, prompting Lyons to fit a touring style front mudguard in place of the abbreviated racing pattern. It was clever foresight and was a contributing factor in his first place in the diabolical conditions that prevailed. His win had the effect of forcing Triumph to somewhat reluctantly market a replica of Lyon's machine. It eventually emerged in 1948 as the Grand Prix; a fully-fledged racer that was available for the private entrant. Manx baker Don Crossley won the 1948 Senior Manx on one and Kiwi Syd Jensen gained a creditable 5th place in the 1949 Senior TT. They proved to be generally unreliable in long-distance events and production was discreetly terminated in 1950.

Following the Manx Grand Prix, Ernie had further success at the prestigious Shelsley Walsh hill climb. He overcame a lack of practice to make FTD at a combined car and motorcycle meeting and in so doing he upset the car brigade considerably. Yet despite Lyons' successes during 1946, Triumph management was implacably opposed to supporting an official works team.

Triumph's negative attitude prompted him to accept an offer to join Nortons for the forthcoming 1947 season, alongside veteran Harold Daniell and new star Artie Bell. Ken Bills was also included in the team, with the understanding that he and Lyons were the junior members. Ernie featured on the leader board during practice for the TT, but race days proved disappointing. He crashed out at the Gooseneck on the opening lap of the Junior and was pronounced unfit to start in the Senior.

At the Belgian Grand Prix on the following weekend he was confident of winning the 350 event and after only one lap his bike blew up. He rode Ken Bills' 350 winner in the 500 race and came in 8th; the first 350 to finish. Apart from a second place in the Senior Ulster Grand Prix behind Artie Bell, it was a generally unproductive year for him on works Nortons.

For the 1948 TT he was entered for the Senior on a Moto Guzzi, sponsored by fellow Irishman Stanley Woods; plus a dealer-entered 7R AJS for the Junior. Lyons' teammate for the Senior was the Italian ace, Omobono Tenni and with the superior speed of the Guzzi both riders were potential winners. Lyons made a meteoric start, only to run out of brakes and crash heavily at the Bungalow on his opening lap. Earlier in the week his AJS fractured a fuel line, which ended his Junior race at the Gooseneck. It had been another disastrous TT week for Ernie Lyons but in the following year he had a remarkable change of fortune.

During the1948-49 off-season, Velocette were burning the midnight oil, designing a new twin-camshaft version of the evergreen KTT. Eight 350s and two 500s were built and these were made available to selected riders.

Nigel Spring acquired a pair of 350s, plus the two 500s, to be ridden by Freddie Frith and Ernie Lyons. Frith took the new model to a win in the Junior TT, after an earlier challenge from AJS stars, Les Graham and Bill Doran. Lyons scored his highest-ever TT finish with a steady second place behind Frith. The 500s were not fancied as potential Senior winners yet once again Ernie Lyons made a leader-board finish with third place behind the Nortons of Harold Daniell and Johnnie Lockett. A major disappointment to Velocette fans was Frith's third lap retirement in what proved to be his final Isle of Man appearance.

Ernie's final road race appearance was the 1950 Junior Ulster, riding a Velo from the Nigel Spring stable. He was well placed in the early stages, battling for the lead with team mate Reg Armstrong and fellow Irishman Louis Carr on a works Norton. The situation was unfolding as a great day for the Irish but Ernie's race ended at Tully Corner, following a tangle with Bill Lomas on another Velocette. Lomas was obviously not impressed with Ernie's actions and relayed his displeasure in no uncertain manner and on that note Ernie retired from the road racing scene. He continued to ride in trials and scrambles until 1963, the year he finally terminated a colourful racing career. Ernie retired to a farm in County Kildare and by happy coincidence his property was sited next door to the farm on which he was born.

Eric Oliver

Eric Oliver

1951 French Grand Prix

Grand Prix sidecar racing attracts a select type of rider; brave and uncompromising. A perfect example was Eric Oliver who gained four world titles from 1949, the initial year of the World Championships until his retirement in 1954. Although he was not officially a 'works' entrant Eric received whole-hearted support from Norton's race shop plus on-going sponsorship from Ron Watson, the creator of the 'Watsonian' brand of sidecars.

Eric Staines Oliver was born in Sussex in 1911 and while still a teenager began a racing career on the grass tracks in the South-East areas of England. Until the outbreak of World War 2 he also rode each year in the Isle of Man. Mostly these efforts ended in retirement until the 1939 series when he was entered on a new MK.VIII KTT Velocette. Eric's perseverance was rewarded with 15th place in both the Junior and Senior TTs.

With the outbreak of war in September 1939 Eric volunteered for service with the RAF. Accepted for aircrew, he went on to fly 40 missions as a flight engineer with Bomber Command. With the end of hostilities Eric resumed his racing activities which included solo and sidecar events. After a successful 1948 season on the European circuits he concentrated on the sidecar category, apart from final solo entries in the 1949 TT series. Eric used Nortons exclusively during his Grand Prix sidecar career and travelled far and wide to compete at the continental race meetings. It was a

Spartan, gypsy existence and any income depended on start and prize money plus bonuses from the oil and tyre suppliers.

The sidecar passengers were also a unique group, heroic characters placing their trust in their drivers' skills. An outstanding example was the bearded journalist, Dennis Jenkinson who passengered for Eric in 1949 and 1950; with the intrepid pair winning the World's Championship in both years. 'Jenks' gained further fame when he navigated for Stirling Moss on his record-breaking drive to win the 1955 Mille Miglia for Mercedes Benz.

Eric's 600cc Norton, with its girder forks and rigid frame looked somewhat dated compared to the much faster Gilera fours ranged against them. It was only Oliver's tigerish riding and 'never say die' attitude that prevailed, particularly on the demanding Continental street circuits. By 1951 the spindly Norton had been replaced by an up-dated 500cc twin-cam version, featuring the latest 'Featherbed' frame. 'Jenks' had also moved on and his place taken by an Italian hero, Lorenzo Dobelli whose presence was a key factor in their gaining a third World Championship for E.S. Oliver.

The following year their meteoric progress was interrupted by a race crash early in the season, where the crew each suffered a broken leg. Undaunted by this setback the durable Oliver turned up for the Belgian Grand Prix with his leg encased in plaster. Dobelli was still on crutches and his situation was taken over by a brave young Belgian. Incredibly, Oliver was allowed to compete but even more so when he snatched victory at the final corner after a stirring battle with the Gileras. Despite these Herculean efforts he suffered mechanical failures at Solitude and Monza and eventually finished in fifth place in the World's Championship. The new champion was Cyril Smith, another Norton exponent. He had suffered a fractured skull earlier in the season and with commendable courage he returned to the fray to become a worthy champion for Great Britain.

By 1953 the BMW influence was creating even greater challenges for the British riders. With its Boxer engine, delivering turbine-like power and shaft drive the 'Rennsport' was ideally suited to sidecar racing and would eventually dominate that category for a further decade. Despite the

BMW's superior performance, Eric Oliver gained his fourth World Championship in 1953. At that stage the single-cylinder Norton had reached the peak of its development, nevertheless the forward-thinking Oliver enhanced its performance with the creation of the 'Kneeler' layout, the styling of which continues to this day.

One of the highlights of the 1954 season was the return of the Sidecar TT to the Isle of Man after an absence of 19 years. Instead of the traditional 37 mile 'Mountain' circuit it was run on the abbreviated Clypse course which used sections of the Mountain circuit but in the reverse direction. It proved to be Oliver's moment of fame; by sheer riding ability and assisted ably by Les Nutt the pair relegated the BMW stars to minor placing. However it was a BMW driven by Wilhem Noll that eventually took the 1954 championship. Norton's glory days had clearly finished and having no desire to switch to a BMW, Eric Oliver retired from the Grand Prix scene. A notable feature of his career was the Belgian Grand Prix where he recorded six consecutive wins between 1949 and 1953.

Following his retirement he opened a motorcycle showroom in Staines, at the same time maintaining an interest in motor sport where he raced a Lotus Elan in local sports car events. However sidecar racing was still an irresistible passion and in 1958 Eric Oliver made a surprise return to the TT. Rather than a pukka racer he was entered on a standard Norton 88 Dominator attached to a Watsonian Monaco touring sidecar. His passenger was Mrs. Pat Wise, herself a racer and following a hectic journey the intrepid pair was placed 10th amongst the purpose-built GP outfits.

In 1960, despite initial doubts from officials the Sidecar TT made a return to the 37 mile Mountain circuit after a lapse of 35 years. Once again E.S. Oliver was an entrant, this time on a GP Norton Kneeler and partnered by Stan Dibben. Supremely confident with their prospects the experienced duo was dealt an almost fatal setback when they were involved in a high-speed practice crash on the Mountain section of the course. The Norton's front forks had snapped off at the steering head, giving them no chance of retrieving the situation. Stan Dibben suffered serious injuries while Eric fared slightly better. Neither of them raced in GP events again, although Eric and Stan did make an appearance at a historic meeting at Brands Hatch in 1978. Sadly time was running out for Eric Oliver and the following year he suffered a severe stroke from which he never recovered. The sidecar maestro was aged 68; a classic example of pure skill and the bulldog spirit, surely the greatest charioteer ever.

John Surtees (1955)

John Surtees

1955 Junior TT

John Surtees' career was notable in that he was the first and in all probability will remain the only racer to achieve a world championship on 2 and 4 wheels. When he was crowned World Driver's Champion at the wheel of a Ferrari in 1964, Surtees already had 7 motorcycle world championships to his name. He is remembered as the complete professional in a demanding sport; one that he pursued in unwavering fashion. His analytical approach to racing was supplemented by thorough mechanical knowledge and exceptional riding ability.

He was born in Surrey in 1934; into a family environment that was steeped in motorcycling. His father was Jack Surtees; a successful sidecar driver in pre-war and early post-war mainland events. Jack was also a motorcycle dealer and one of the first agents for Vincent motorcycles in southeast England.

John's road racing career began modestly enough on a 250 Triumph fitted with McCandless rear suspension. The Tiger 70 was a pre-war design, on which he competed regularly at Brands Hatch and other venues. At the time he was serving an apprenticeship with the Vincent firm at Stevenage, and was able to purchase one of their Grey Flash racers, which came as a collection of parts; and with typical thoroughness John re-built it to better than work's specification.

Throughout 1951 and into 1952 the Vincent was raced successfully at a variety of British short circuits. At the peak of its development it was a rare occasion for the Surtees-Vincent combination to be beaten. Despite his success, John was well aware that in order to remain competitive he would need to graduate to a more sophisticated racer.

The Featherbed Manx Norton was becoming more readily available to the private owner, provided his credentials satisfied Norton's management. Their managing director, Gilbert Smith was prepared to supply a new 500 Manx to John Surtees, with a proviso that he competed in international events. It placed John in an awkward situation; for he intended to ride in the amateur Manx Grand Prix to gain experience in the Isle of Man. As a result he had to forgo the Manx and make his international debut in the 1952 Senior Ulster Grand Prix.

The Ulster was a memorable experience for Surtees, for it was the final year that the race was run on the old Clady circuit, before it was moved to Dundrod. He had the opportunity to compete against legendary riders, such as Les Graham and Cromie McCandless on that demanding circuit. It was also an historic outing for McCandless, making his debut aboard a works Gilera. He scored a resounding win against fierce opposition; and on that note the Ulster star retired from road racing.

Following the Ulster, which resulted in a 6th place for Surtees, he campaigned the new Norton at national-level meetings in England. In early 1953 he added a second hand 350 Manx Norton to his stable This example had a chequered history; being the machine that Ray Amm used before his elevation to 'works' status. During the latter part of 1952 and the 1953 season the Surtees Nortons were entered in sixty races. It was a strenuous schedule, and a measure of John's ability was that he won thirty of these events.

In June 1953 he made his Isle of Man debut. Initially he was to ride his own Nortons, plus an entry of a 125 EMC. During practice the Norton works team found itself short of a rider following Syd Lawton's crash at Brandish. Joe Craig paid John a rare compliment with an invitation to join the team, despite not having previously ridden at the TT. Craig expressed his reservations when John informed him that he still intended to ride the EMC in the 125 Lightweight event.

Craig was even less pleased when John crashed the EMC at Ballaugh Bridge on his first practice lap and broke a bone in his wrist. It was a disastrous TT for Surtees, having to forgo his opportunity of a works ride. Craig later confided that he did consider having him in the 1954 team, and it was 1955 before Craig finally relented and Surtees was included. Their 1955 team comprised John Surtees John Hartle and veteran Jack Brett. For a works team it was a very low-key effort, and rather than purpose-built racers, they used development versions of the production Manx Norton. The riders' salary was limited to a 500 pounds retainer from Castrol, which was quite a paltry amount, even for those rather impecunious times.

Apart from occasional works appearances Surtees rode his own Nortons, plus a very competitive 250 NSU. His start/win ratio of wins at National meetings was quite remarkable, and far too numerous to record in this volume. He also had the opportunity to ride a works prepared BMW at the Nurburgring, and despite his unfamiliarity with the bike and the circuit he put in a most creditable performance. Surtees was impressed enough with the Munich twin to be confident of winning a World's Championship on one. However the BMW directors seemed unable to come to any decision whether or not to engage him. When they finally overcame their inertia and made an approach, Surtees had already signed with MV.

The move proved to be most advantageous for both the MV concern and for Surtees, although it was never easy for a foreigner to be accepted into the inner sanctum of an Italian organization. Les Graham had joined MV in 1951 and with a combination of his technical expertise and riding skills he turned the ungainly four into a potential world-beater with end of season wins in the 1952 Spanish and

Italian GPs. Graham preferred to use the unwieldy Earles forks on his machines, which may have been a factor in his fatal crash at the 1953 Senior TT.

Ray Amm's move to MV in 1955 was expected to elevate their situation and once again fate stepped in during Amm's debut at Imola. John Surtees' analytical approach to the development of the current machines, plus his undoubted riding ability was vindicated in his very first season with them. His TT debut on the MV resulted in a convincing Senior win, ahead of the Nortons of John Hartle and Jack Brett and only for running out of fuel in the Junior he could well have made it a double. Surtees went on to score emphatic wins at the 500 Dutch TT and 500 Belgian Grand Prix. These victories, plus his TT win gave him a substantial points lead in the 500cc World Championship.

Surtees had the misfortune to crash during the 350 race at the Solitude circuit outside Stuttgart. He suffered a broken right arm in the accident which sidelined him until practically the end of the season. He returned to score a win at the 500 Spanish Grand Prix. This was a non-championship event and Surtees had amassed sufficient points to consolidate his world championship lead.

A factor in his win was the absence of Geoff Duke; for the first half of the season at least, when Duke had the misfortune to be stripped of his international licence for a six-month period. This wretched affair had its origins at the 1955 Dutch TT at Assen. With something like 120,000 paying spectators the event was a financial bonanza for the organizers. However they baulked at increasing the miserable amount of start money they were offering the private entrants. Those battling individuals who make up the bulk of the field are under considerable expense with travel costs, plus having to provide their own machinery. When their appeal for more start money was rejected, the riders threatened to retire en masse at the conclusion of the opening lap of the Junior TT, The organisers considered they were bluffing and still refused to concede to their demands. As a last resort the privateers appealed to the works riders to use their influence. Duke, Armstrong and several other works riders offered to support their cause. They were the sole representatives and the ramifications of the affair proved to be far-reaching and quite expensive for them.

Pandemonium erupted at the end of lap one of the Junior TT when the bulk of the private entrants retired, leaving the works riders to continue. The threat of a repeat performance in the Senior event caused the organizers to finally relent and pay up. It was a distasteful episode and the governing body of motorcycle sport, the F.I.M. was out for revenge. Their venom was directed at Duke and 12 others and in a most vindictive move their competition licences were withdrawn. This had the effect of scuttling Duke's 1956 championship aspirations. He did make a return during the season and by then as they say the horse had bolted.

Historians would probably agree that 1957 represented a year of unparalleled technical interest. With Duke reinstated he became a serious contender for 500cc honours for Gilera. His teammate Liberati was another possibility, although he opted not to ride at the Isle of Man. Moto Guzzi had at last achieved an element of reliability with their fabulous water-cooled V8 and were not to be discounted either. Zeller's BMW had the reliability for top honours but could be lacking in outright speed. The stage was set for the most exciting season ever.

Reigning world champion John Surtees had recovered well from his injury and was facing the New Year with optimism. As events turned out, his machines were plagued with persistent breakdowns that saw him finish third in the 500cc championship.

Duke's campaign began in disastrous fashion. Imola was an early event in the racing calendar where Duke suffered a broken collarbone after stepping off his Gilera in the 500 event. He did not re-appear until late in the season and by then any hope of another world championship had vanished. Thus, at the end of 1957 two new names were added to the list of previous champions; the effervescent Aussie, Keith Campbell with the 350 crown for Moto Guzzi and the brooding, handsome Italian, Libero Liberati the 500 for Gilera..

Meanwhile the dramas of 1957 were unfinished. At season's end came the bombshell that the leading Italian manufacturers were withdrawing from racing. These included Gilera, MV Agusta, Moto Guzzi and Mondial; due, they claimed to the unsupportable costs. It was shattering news, and in many instances it spelt the end of riders' careers.

With its rivals committed to leaving the scene, MV reversed its decision and in 1958 they began a full-blooded assault on the world championships. Their efforts were concentrated on every category from 125 to 500cc and in this they were unchallenged. In the 500 class it was a pattern that continued for the next decade and from 1958 to 1960, when he announced his retirement from motorcycle competition,

John Surtees won the Junior/Senior crown in every year. His dominance was absolute, which made life exceedingly difficult for the battling privateers who made up the field on Manx Nortons or Matchless G50s.

With Surtees' move to Formula One the 500cc mantle was passed to other MV greats, beginning with Rhodesian Gary Hocking and Britain's Mike Hailwood. MV's favourite champion was surely their local hero Giacomo Agostini who brought them a multitude of World Championships on the Gallarate 'fire engines'. It fell to the brilliant but abrasive Phil Read to give MV their last taste of glory during the 1974 GP season. By then the Japanese influence was indisputable. Of all the champions none could match the all-round brilliance of John Surtees, who with singular dedication made that transition from world champion on two wheels to world champion on four.

He still delights enthusiasts with his scintillating displays at historic car and bike meetings.

Author's Tribute

This concludes a review of a bygone era; sportsmen of diverse character and outlook. Icons of their chosen profession, bold and courageous they ventured into a realm of excitement that only a chosen few could negotiate. Talented as they all were, various names rise above the pack. In a pre-war era one must favour a select few. Guthrie, Woods and Frith are names that spring to mind. While in a post-war situation, Duke, Surtees and Bell evoke similar emotions. Just a handful of those maestros are still with us today and one thing is certain; we may never see their like again.

Murray McLeod©
Central Coast NSW
2012

A TOUCH OF NOSTALGIA

FROM THE AUTHOR'S ARCHIVES

1948 MSS VELOCETTE   1949 MATCHLESS G80C

1951 VINCENT COMET   1963 NORTON ATLAS